Springer Theses

Recognizing Outstanding Ph.D. Research

For further volumes:
http://www.springer.com/series/8790

Aims and Scope

The series "Springer Theses" brings together a selection of the very best Ph.D. theses from around the world and across the physical sciences. Nominated and endorsed by two recognized specialists, each published volume has been selected for its scientific excellence and the high impact of its contents for the pertinent field of research. For greater accessibility to non-specialists, the published versions include an extended introduction, as well as a foreword by the student's supervisor explaining the special relevance of the work for the field. As a whole, the series will provide a valuable resource both for newcomers to the research fields described, and for other scientists seeking detailed background information on special questions. Finally, it provides an accredited documentation of the valuable contributions made by today's younger generation of scientists.

Theses are accepted into the series by invited nomination only and must fulfill all of the following criteria

- They must be written in good English.
- The topic should fall within the confines of Chemistry, Physics and related interdisciplinary fields such as Materials, Nanoscience, Chemical Engineering, Complex Systems and Biophysics.
- The work reported in the thesis must represent a significant scientific advance.
- If the thesis includes previously published material, permission to reproduce this must be gained from the respective copyright holder.
- They must have been examined and passed during the 12 months prior to nomination.
- Each thesis should include a foreword by the supervisor outlining the significance of its content.
- The theses should have a clearly defined structure including an introduction accessible to scientists not expert in that particular field.

Alexander V. Yakubovich

Theory of Phase Transitions in Polypeptides and Proteins

Doctoral Thesis accepted by
The Goethe University of Frankfurt, Germany

 Springer

Author
Dr. Alexander V. Yakubovich
Frankfurt Institute for Advanced Studies
Goethe University
Ruth-Moufang-Str. 1
60438 Frankfurt am Main
Germany
e-mail: yakubovich@fias.uni-frankfurt.de

Supervisor
Prof. Dr. Andrey Solov'yov
Frankfurt Institute for Advanced Studies
Goethe University
Ruth-Moufang-Str. 1
60438 Frankfurt am Main
Germany
e-mail: solovyov@fias.uni-frankfurt.de

ISSN 2190-5053
ISBN 978-3-642-26953-0
DOI 10.1007/978-3-642-22592-5
Springer Heidelberg Dordrecht London New York

e-ISSN 2190-5061
ISBN 978-3-642-22592-5 (eBook)

Cover design: eStudio Calamar, Berlin/Figueres

Printed on acid-free paper

Springer is part of Springer Science+Business Media (www.springer.com)

Supervisor's Foreword

The important goal of numerous current investigations is to generate new, detailed knowledge about the nanoscale mechanisms leading to global conformational changes of single biomacromolecules. This goal is being tackled by different theoretical and experimental approaches and methods. In foreword to this thesis I want to emphasize the key theoretical framework which is based on an inter-disciplinary approach combining the methods of molecular dynamics, statistical mechanics, computational chemical physics, and quantum mechanics aiming to provide a comprehensive description of phase transitions and cooperative dynamics in peptides, proteins, and other biomacromolecules. Understanding such structural transformations reveals a tremendous amount of useful information about the properties of these systems, including how they function and how they are regulated. These transitions generally correspond to the finite system analogue of a phase transition.

There have been numerous experimental studies of polypeptide and protein folding and general structural transitions in proteins and biomacromolecular complexes. In spite of this effort the dynamics of most of these phenomena is not well understood. This Ph.D. thesis attempts to improve this situation and to advance our understanding of biomolecular dynamics within the framework of statistical mechanics and phase transition theory. This research is particularly timely because experiments now allow one to monitor the dynamics of single biomolecules.

Building a comprehensive understanding of phase transition-like phenomena in finite systems is a challenging problem, with a variety of applications in biophysics and nanosystems, including protein folding and misfolding. This Ph.D. thesis develops a general theoretical framework for the description of phase transitions in proteins, peptides, protein domains and other biomacromolecules. The unifying basis for the description is analysis of the potential energy surface (PES).

Phase transitions have been an active research field for more than a century, and there is currently great interest in the cooperative dynamics and phase transition-like processes in *finite systems*. For structural transitions in complex biomacro-molecules, neither an analytical solution nor a brute force numerical computation are feasible.

Even for the most advanced computers molecular dynamics simulations are limited to a microsecond timescale for relatively small proteins. Furthermore, such runs generally do not provide adequate statistics for a proper sampling. Alternative approximations are clearly required. Statistical mechanics provides a mature framework for dealing with such processes. It defines the partition function, which allows one to construct a parameter-free description of the observable properties of a system.

Establishing fundamental connections between the statistical mechanics methods for calculating partition functions with the modern computational techniques for molecular dynamics is one of the very promising research directions, which thoroughly explored in the thesis and brought new important insights to the old standing problem of protein folding. Thus, it was demonstrated that the combined statistical mechanics and molecular dynamics methods provide a very useful tool for quantitative description of phase transitions and cooperative changes in large biomacromolecules. The theoretical analysis performed in the thesis has been validated by comparison with experiment.

The earliest attempts to describe phase transitions in polypeptide chains within the framework of statistical physics goes back to the late 1950s, when a general formalism for the construction of the partition function of polypeptides was suggested. This early work considered the α-helix $\leftrightarrow$ random coil phase transition in polypeptides as a model to analyse conformational changes in globular proteins.

A common feature of previous and current theories is that various parameters (such as enthalpy, entropy, and free energy changes) enter the partition function. These parameters can be deduced from analysis of available experimental data or from theoretical calculations. For example, the parameters of the Zimm–Bragg theory can be determined from optical circular dichroism measurements. The first attempts to evaluate these parameters theoretically date to early 1970s. These studies used semi-empirical potentials to describe the conformational dynamics of a polypeptide, and determine the temperature of the helix–coil transition in a polypeptide chain.

In principle, all the necessary information for the construction of a partition function can be obtained from the PES, which for simpler systems, such as peptides, can be calculated even at a quantum mechanical level. Thus, knowledge of the stationary points of the PES is sufficient for the construction of the partition function of a biomolecule, and thus provides a complete thermodynamic description, which includes the calculation of all the essential thermodynamic variables and characteristics, including the heat capacity, phase transition temperature, and free energy.

For large biomolecular systems the fundamental problem with applying standard techniques, such as molecular dynamics and Monte Carlo simulations, is that the phase space is too large, so that converged results cannot be obtained in a reasonable time. Basing theoretical analysis on stationary points of the PES one immediately introduces some coarse-graining. It is then necessary to express observable properties, both thermodynamic and kinetic, in terms of the distribution of stationary points and their intrinsic characteristics. This analysis can be carried out in a formally exact manner, for example, writing the global partition function in terms of a sum over local minima. Hence there exist well-defined limits where the exact result can be derived. In practice, one wishes to obtain results of useful accuracy within a much smaller amount of computer time, and this can be achieved by a series of well-defined approximations.

Coarse-graining the system by considering stationary points instead of the full phase space achieves an immediate simplification in calculations of both thermodynamic and kinetic properties. Instead of using the whole phase space one can construct model densities of states for the stationary points, i.e. local partition functions, and combine these to calculate global properties. Unfortunately, the number of stationary points generally grows exponentially with system size, so a complete enumeration is only possible for relatively small systems. One must therefore sample the required stationary points according to an appropriate distribution. Analyses of the stationary points aiming to identify the essential degrees of freedom enables the individual partition functions in the superposition sum to be factorised and simplified.

Analysis of the PES has previously been used to describe folding, conformational changes, and dissociation transitions in various peptides. More recent applications of discrete path sampling involve protein misfolding and analysis of large-scale conformational transformations associated with function. In each case it was possible to obtain rate constants that compare well with the experimental values, and make detailed predictions about the mechanisms involved at an atomistic level of detail.

It is important to note that the connection between the PES and a statistical description is rather general. Coarse-graining the PES in terms of stationary points and their connections can be used to describe phase transitions and cooperative dynamics in a variety of complex biomolecular and finite size systems. One recent application involves melting of fullerenes, and there have been numerous applications to other atomic and molecular clusters.

In summary, this thesis presents the results of a systematic comprehensive study of polypeptide and protein folding processes based on a mature set of theoretical tools that provide detailed insight via computation into the structure, dynamics and thermodynamics of biomacromolecules. This theoretical foundation is based on the statistical mechanics of the PES, global optimisation techniques and molecular dynamics. Detailed comparison with experiments is considered for phase transitions and cooperative structural changes of biomacromolecules, including polypeptides and globular proteins. In each case changes in structure are characterised in the context of phase transition theory for finite systems, and both the thermodynamic functions are calculated and compared with each other and with experiment. A firm theoretical foundation for the phase transition description is provided by a rigorous mathematical analysis of the connection between stationary points of the PES, the appearance of a phase transition, and the dynamics of folding.

Frankfurt am Main, August 2011   Prof. Dr. Andrey Solov'yov

Acknowledgments

The research which is reported has been conducted during the years 2006–2010 while employed at the Frankfurt Institute for Advanced Studies (FIAS). First and foremost I thank Prof. Dr. Andrey Solov'yov and Prof. Dr. h. c. mult. Walter Greiner for the uncounted enlightening talks we had during this time. The strong interest in the progress of this work that they took at all times. Their valuable comments and constructive criticism, expressed in a straightforward but friendly manner, not only helped me gain new insights but also motivated me time and time again. I am glad to have worked with them and grateful for having been their student.

While working on my dissertation, I profited a lot from the many opportunities to present and discuss my work-in-progress in various meetings at the Frankfurt Institute for Advanced Studies. My special gratitude goes to the members of the Meso-Bio-Nano Science Group, especially to Dr. Andrey Korol and Dr. Elsa Henriques.

My special acknowledgments goes to Dr. Ilia Solov'yov, in a close collaboration with whom we did a substantial part of the present work. I would also like to thank him, Ms. Elena Solov'yova, Mr. Marc Henniges and Prof. Dr. Stefan Schramm for helping me with the German Zusammenfassung. I am very greatful to Dr. Eugene Surdutovich for his help in the proofreading this manuscript. The possibility to perform complex computer simulations at the Frankfurt Center for Scientific Computing is gratefully acknowledged.

Most importantly, none of this would have been possible without the love and patience of my family. Especially I want to mention my parents Valentine and Elena, my sister Olga, my grandmothers Valentina and Irina, my grandfathers Ivan and Igor, my aunt Svetlana and my uncles Stanislav and Alexander, my cousins Mikhail, Kira, Igor and Ivan. My family has been a constant source of love, concern, support and strength all these years. I would like to express my heart-felt gratitude to my family.

Contents

Chapter 1
Introduction

A protein is a polypeptide chain consisting of a sequence of units or "residues", which are amino acids chosen from a pool of 20. Proteins are synthesized as unfolded polypeptide chains and they fold after synthesis in order to become active. Anfinsen [1] realized that the driving force for folding is the gradient of free energy and the search for the free energy minimum gives the three-dimensional (3D) structure, which is the most stable structure.

Protein folding refers to the process by which a protein assumes its characteristic structure, known as the native state. The most fundamental question of how an amino acid sequence specifies both a native structure and the pathway to attain that state has defined the protein folding field. Over more than four decades the protein folding field has evolved [2], as have the questions pertaining to it.

Proteins are involved in virtually every biological process in a living system. Therefore there is enormous number of possible biological and medical applications of proteins in living organisms. The ultimate goal of the modern chemical engineering and protein design science is to propose an amino acid sequence with specific structure and function for each particular application. The inverse problem to these task is to predict the structure of a protein with a given amino acid sequence. The protein structure prediction problem is a fundamental problem treated across disciplines. Many approaches to computational protein structure prediction using first principles have been developed over the last decade that are based on Anfinsen's thermodynamic hypothesis. Computational structure prediction based on first principles is, however, not the only way to determine protein structure. The number of protein structures that have been determined experimentally continues to grow rapidly. At the end of 2009, the number of structures freely available from the protein data bank [3] is approaching 60,000. The availability of experimental data on protein structures has inspired the development of methods for computational structure prediction that are knowledge-based rather than physics based. In contrast to methods that attempt to minimize the free energy and derive the structure from first principles, these knowledge-based approaches search databases of known structures to infer information about an amino acid sequence of unknown 3D structure. Nowadays, these knowlegde-based methods are the most successful in protein structure

A. V. Yakubovich, *Theory of Phase Transitions in Polypeptides and Proteins*, Springer Theses, DOI: 10.1007/978-3-642-22592-5_1,
© Springer-Verlag Berlin Heidelberg 2011

prediction. However, such database methods have been criticized for not helping to obtain a fundamental understanding of the mechanisms that drive structure formation [4]. The aim of these thesis is to investigate the fundamental driving forces of the folding$\leftrightarrow$ unfolding transition and to provide the statistical mechanics model that treats the folding$\leftrightarrow$ unfolding transition form the point of view of statistical mechanics. The description of the system in terms of equilibrium thermodynamics allows one to derive the thermodynamical properties of the system on the timescales that are not feasible in any molecular dynamics simulations. Another advantage of statistical mechanics is that this approach provides a "transparent" physical picture of the fundamental forces and interactions in the system, in contrast to molecular dynamics simulations, where the final result often lacks the complete understanding.

In simple terms, folding could be described as the process by which many degrees of freedom existing in unfolded polypeptide chains become coordinated into well-defined structures through energetics specific to their amino acid sequences. Protein structures are defined by thousands of atomic coordinates; therefore even if we ignored the surrounding solvent molecules, it would still be impossible to discern which of the astronomical number of possible conformations are physically relevant. Furthermore, protein structures are marginally stabilized by dense networks of weak noncovalent interactions, so that the smallest imprecision in calculating protein energetics leads to large relative errors. In other words, the understanding of protein folding is constrained by limitations in sampling and in the intrinsic simplifications of the procedures used to correlate energy with conformation [2].

Francis Crick [5] wrote about the challenge of the protein folding problem:

> Nature performs these folding calculations effortlessly, accurately, and in parallel, a combination we cannot hope to imitate exactly. Moreover, evolution will have found good strategies for exploring many of the possible structures in such a way that shortcuts can be taken on the path to the correct fold. The final structure is a delicate balance between two numbers, the energy of attraction between the atoms, and the energy of repulsion. Each of these is very difficult to calculate accurately, yet to estimate the free energy of any structure we have to estimate their difference. The fact that it usually happens in aqueous solution, so that we have to allow for many water molecules bordering the protein, makes the problem even more difficult.

A protein can be described at many levels. At the finest level, one would simply treat the entire system with all the degrees of freedom with the laws of quantum mechanics. The difficulties associated with a first-principles quantum mechanical approach include the large number of degrees of freedom; the necessity of calculating the interactions during the dynamical process of folding, with the solvent taken into account in an accurate manner; and, even if the interactions were known exactly, the limitations of present-day computers in accurately following the dynamics through the folding process. Simulating such a system at this level of description is a daunting task and has not yet been achieved. More fundamentally, such an approach would enable one to mimic nature but not necessarily understand her.

A more practical approach is to define a small number of degrees of freedom that describe the coarse features of the protein solvent complex, thereby reducing the hyperdimensional potential energy surface to a much simplified potential energy surface of low dimensionality. For the efficient description of the folding↔ unfolding transition one has to accurately determine all the principal degrees of freedom the are responsible for the conformational transitions in a biomolecule. Further simplification of the description of the statistical mechanics properties of a polypeptide or a protein can be achieved if one can distinguish only the statistically significant domains of the potential energy surface of the system of reduced dimensionality. This is not a trivial task of an arbitrary biomolecule. However, when such domains on the potential energy surface of reduced dimensionality are determined, one can obtain all the thermodynamical properties of the system. This procedure effectively connects the worlds of theory, computer simulation, and experiment in protein folding. Low-dimensional potential energy projections provide tools to condense the wealth of structural and dynamic data generated in large-scale molecular simulations [6–9] and to analyze quantitatively the data obtained in protein folding experiments [10–12]. Nevertheless, connecting the worlds of theoretical prediction and empirical observation (both in vitro and *in silica*) comes at an expense [13].

1.1 Problems Addressed in the Thesis

The aim of the thesis is to provide a theoretical model for the description of the process of polypeptide and protein folding. The major challenge of this work is to converge the theoretical description of the folding process performed with the methods solely based on fundamental physical principles with the experimental measurements of protein folding in vitro. In order to achieve this goal the following problems were addressed:

1. The potential energy surfaces of small fragments of proteins, polypeptides, consisting of several amino acids were calculated using the ab initio methods of quantum mechanics. The results of these calculations are reported in Refs. [14–18] and discussed in Chap. 3.
2. Conformational transitions in polypeptides and proteins can be understood as a phase transitions and treated with the methods of statistical physics. Knowing the potential energy surface of the system one can construct its partition function and derive all thermodynamics functions of the system. The formalism of the construction of the partition function for the polypeptides in the gas phase is outlined in Refs. [19, 20] an in Chap. 4 of the thesis.
3. In order to benchmark the accuracy of the developed in Refs. [19, 20] statistical mechanics formalism it is necessary to compare the results of the statistical mechanics model with the results of molecular dynamics simulations. Unfortunately, currently there are no experimental measurements of the thermodynamic properties of polypeptides in the gas phase. But the properties of single biomolecules in the gas phase are nowadays intensively investigated [21–23]. The thorough comparison of the results of statistical mechanics model with the

results of molecular dynamics simulation is performed in Chap. 5 and in Refs.
[24].

4. Using molecular dynamics simulations it is possible to investigate the conformational transitions in various polypeptides. In Sect. 5.4 of the thesis and in Ref.
   [25] the results of molecular dynamics simulations of conformational transitions
   in alanine, valine and leucine polypeptides are presented. The analysis of the
   molecular dynamics simulations is accompanied by a discussion of the effective
   ways of obtaining the thermodynamic functions of the system, in particular heat
   capacity on temperature dependence from the molecular dynamics simulations.

5. For the description of thermodynamic properties of polypeptides and proteins in
   aqueous environment it is necessary to account for solvent effects. In Ref. [26] is
   presented a way for the construction of the partition function of the polypeptide
   in water solution.

6. In Ref. [27] and in Chap. 6 the way of the construction of partition function of the
   protein in water environment is presented and the comparison of the predictions
   of the statistical mechanics model with the results of experimental measurements
   of the heat capacity on temperature dependencies is performed for two globular proteins, staphylococcal nuclease and metmyoglobin. The comparison of the
   results of the statistical mechanics model with the direct experimental measurements of the heat capacity allows one to conclude about the accuracy and the
   range of applicability of the developed theoretical formalism.

All the aforementioned problems are discussed in detail in the thesis. I hope that
this work will provide one more bridge between the very intriguing and long standing
interdisciplinary problem of protein folding and the deterministic world of theoretical
physics.

The thesis is structured as follows. In Chap. 2 is presented an overview of the
methods of quantum mechanics which are used for the ab initio calculations of
potential energy surfaces of short alanine and glycine polypeptides. In Chap. 3 are
presented the results of calculations of the potential energy surfaces of alanine and
glycine polypeptides as functions of the dihedral angles φ, ψ and ω. In Chap. 3 is
performed the analysis of the potential energy surfaces and discussed the transitions
between different conformational states of short polypeptides. In Chap. 4 the partition function of a polypeptide is derived. The helix $\leftrightarrow$ coil conformational transition
in a polypeptide is considered as a phase transition in a finite system. The discussion of the comparison of the results of the developed statistical mechanics model
with the results of molecular dynamics simulations of conformational transitions
in alanine polypeptides of different length is presented in Chap. 5. The conformational transitions in valine and leucine polypeptides are discussed in Sect. 5.4 of the
thesis. The partition function of a single-domain protein in water environment is
derived in Chap. 6. In Sect. 6.3 the results of the statistical mechanics model for the
description of conformational transitions in proteins are compared with the results of
experimental measurements of heat capacity on temperature dependence for staphylococcal nuclease and metmyoglobin. Chapter 7 presents the summary of the results
of the thesis and conclusions.

References

1. Anfinsen, C. (1973). Principles that govern the folding of protein chains. *Science, 181*, 223–230.
2. Banavar, J., & Maritan, A. (2007). Physics of proteins. *Annual Review of Biophysics & Biomolecular Structure, 36*, 261–280.
3. Protein data bank. (2010). http://www.rcsb.org
4. Floudas, C., Fung, H., McAllister, S., Monnigmann, M., & Rajgaria, R. (2006). Advances in protein structure prediction and de novo protein design: a review. *Chemical Engineering Science, 61*, 966–988.
5. Crick, F. (1988). *What mad pursuit: a personal view of science*. New York: Basic Books.
6. Boczko, E., & Brooks, C. (1995). First-principles calculation of the folding free-energy of a 3-helix bundle protein. *Science 296*, 393–396.
7. Camacho, C., & Thirumalai, D. (1993). Kinetics and thermodynamics of folding in model proteins. *Proceedings of the national academy of sciences of the United States of America 90*, 6369–6372.
8. Garcia, A., & Onuchic, J. (2003). Folding a protein in a computer: an atomic description of the folding/unfolding of protein. *Proceedings of the national academy of sciences of the United States of America, 100*, 13898–13903.
9. Pande, V., Baker, I., Chapman, J., Elmer, S., & Khalig, S. (2003). Atomistic protein folding simulations on the submillisecond time scale using worldwide distributed computing. *Biopolymers 68*, 91–109.
10. Alm, E., & Baker, D. (1999). Physics of proteins. *Proceedings of the national academy of sciences of the United States of America, 96*, 11305–11310.
11. Henry, E., & Eaton, W. (2004). Combinatorial modeling of protein folding kinetics: free energy profiles and rates. *Chemical Physics, 307*, 163–185.
12. Munoz, V. (2002). Thermodynamics and kinetics of downhill protein folding investigated with a simple statistical mechanical model. *International Journal of Quantum Chemistry, 90*, 1522–1528.
13. Muñoz, V. (2007). Conformational dynamics and ensembles in protein folding. *Annual Review of Biophysics & Biomolecular Structure, 36*, 395–412.
14. Yakubovich A., Solov'yov I., Solov'yov A., & Greiner W. (2006). Conformational changes in glycine tri- and hexapeptide. *The European Physical Journal D, 39*, 23–34.
15. Yakubovich, A., Solov'yov, I., Solov'yov, A., & Greiner, W. (2006). Conformations of glycine polypeptides. *Khimicheskaya Fizika (Chemical Physics), 25*, 11–23(in Russian).
16. Solov'yov, I., Yakubovich, A., Solov'yov, A., & Greiner, W. (2006). On the fragmentation of biomolecules: fragmentation of alanine dipeptide along the polypeptide chain. *The Journal of Experimental and Theoretical Physics, 103*, 463–471.
17. Solov'yov, I., Yakubovich, A., Solov'yov, A., & Greiner, W. (2006). Ab initio study of alanine polypeptide chain twisting. *Physical Review E, 73*(1–10), 021916.
18. Solov'yov, I., Yakubovich, A., Solov'yov, A., & Greiner, A. (2006). Potential energy surface for alanine polypeptide chains. *Journal of Experimental and Theoretical Physics, 102*, 314–326.
19. Yakubovich, A., Solov'yov, I., Solov'yov, A., & Greiner, W. (2006). Phase transition in polypeptides: a step towards the understanding of protein folding. *European Physical Journal D, 40*, 363–367.
20. Yakubovich, A., Solov'yov, I., Solov'yov, A., & Greiner, W. (2007). Ab initio description of phase transitions in finite bio- nano-systems. *Europhysics News, 38*, 10–10.
21. Andersen, L., & Bochenkova, A. (2009). The photophysics of isolated protein chromophores. *European Physical Journal D, 51*, 5–14.
22. Shintake, T. (2008). Possibility of single biomolecule imaging with coherent amplification of weak scattering x-ray photons. *Physical Review A, 78*(1–9), 041906.
23. Pazera, T., & Sachs, A., (Eds.). (2009). DESY 2008. Wissenschaftlicher Jahresbericht des Forschungszentrums DESY. Deutsches Elektrinen-Synchrotron DESY.

24. Solov'yov, I., Yakubovich, A., Solov'yov, A., & Greiner, W. (2008) α-helix$\leftrightarrow$ random coil phase transition: analysis of ab initio theory predictions. *European Physical Journal D, 46,* 227–240.
25. Yakubovich, A., Solov'yov, I., Solov'yov, A., & Greiner, W. (2008). Phase transitions in polypeptides: analysis of energy fluctuations. *European Physical Journal D, 51,* 25–31.
26. Yakubovich, A., Solov'yov, A., & Greiner, W. (2010). Conformational changes in polypeptides and proteins. *International Journal of Quantum Chemistry, 110,* 257–269.
27. Yakubovich, A., Solov'yov, A., & Greiner, W. (2009). Statistical mechanics model for protein folding. *AIP Conference Proceedings, 1197,* 186–200.

Chapter 2
Theoretical Methods of Quantum Mechanics

2.1 Introduction

Biochemical processes occur on different scales of length and time [1] ranging from a few angstroms, the size of the active site of proteins, where the ultrafast triggering steps usually take place, up to the level of the cells and organs, where their macroscopic effects are detectable by the naked eye. Intermediate steps are the structural rearrangement of biomolecules (approximately nanometer and 10–100 ns), their aggregation/separation and folding/unfolding (10 nm to micrometer and greater than microsecond) and internal cell diffusion and dynamics (micrometers to millimeters and milliseconds to hours). This inherent hierarchical organization is responsible for the complexity of living matter: a single process involves a multiscale cascade of events whose description requires the combination of different methodologies in so-called multiscale approaches [2, 3].

At any resolution, the quality of a model depends on the accuracy with which the two following issues are addressed: the description of the interactions and the sampling of the configurations of the system. In this respect, there are a few concepts that iteratively occur.

First concept regards the potential energy surface. The method used to evaluate the potential energy surface strictly depends on the resolution level. For small molecules (up to a few tens of atoms), both the nuclear and electronic degrees of freedom can and must be explicitly treated in order to describe the electronic structure of the molecule. The concept of the potential energy surface is related to the Born–Oppenheimer approximation [4], assuming that the much faster electrons adiabatically adjust their motion to that of the atomic nuclei. Thus, at any time, the Schrödinger equation for the electron system is to be solved in the external field generated by the atomic nuclei considered as frozen, and one is left with a nuclear-configuration dependent set of energy eigenvalues $E_i(\{R_i\})$ that define the potential energy surface of the ground and excited states. In turn, the potential energy surfaces are effective electronic structure-dependent potential energy functions that determine the dynamics of the nuclei.

A. V. Yakubovich, *Theory of Phase Transitions in Polypeptides and Proteins*,
Springer Theses, DOI: 10.1007/978-3-642-22592-5_2,
© Springer-Verlag Berlin Heidelberg 2011

The different methods used to solve the Schrödinger equation, called quantum mechanics approaches, basically differ by the way the electron–electron interactions are treated. The electron correlation can be accurately added as a perturbation of the exchange-only Hartree–Fock (HF) scheme at the expense of a large computational cost or via less expensive (and less predictive) semiempirical Hamiltonians [5]. Alternatively, in density functional theory (DFT) and, more specifically, in the Kohn–Sham scheme, the many-electron problem is reduced to a single-electron Schrödinger problem in a self-consistent exchange-correlation potential depending on the electron density. DFT changed the way of approaching the quantum mechanics calculations: its accuracy and predictive power are comparable to those of other ab initio methods but much cheaper computationally. Thus, density functional theory is conveniently used for molecular structure optimization or even for dynamic exploration of the potential energy surfaces, that is, ab initio molecular dynamics. The density functional theory is also intensively used for the derivation of parameters of molecular mechanics potentials (See. Sect. 4.2). In addition, excited-state calculations are possible with the time-dependent extension of DFT [6]. Although time-dependent DFT is known to suffer from large errors in the excitation energies, it is often used in biosystems thanks to its extremely low cost with respect to other excited-state methods.

The following sections of this chapter is devoted to the brief description of the methods of quantum mechanics that are used in Chap. 3 for the calculations of potential energy surfaces of the polypeptides.

2.2 The Schrödinger Equation

For exact description of the electronic and ionic structure of a multi atomic system one has to solve the Schrödinger equation for all particles in the system.

The Schrödinger equation describes the wavefunction of the system (see e.g. [7]):

$$\hat{H}\Psi(\mathbf{r}, \mathbf{R}, t) = i\frac{\partial \Psi(\mathbf{r}, \mathbf{R}, t)}{\partial t}, \tag{2.1}$$

where $\hat{H}$ is the Hamilton operator (Hamiltonian), $\Psi(\mathbf{r}, \mathbf{R}, t)$ is the wavefunction of the system, which depends on the coordinates of the electrons and the nuclei within the system, and time. Let us designate them as $\mathbf{r}$, $\mathbf{R}$ and t, respectively. In this section the atomic system of units is used, $\hbar = m_e = |e| = 1$ unless other units are not indicated.

The Hamiltonian is a sum of kinetic, $\hat{T}$, and potential, $\hat{V}$, energy terms:

$$\hat{H} = \hat{T} + \hat{V} \tag{2.2}$$

If $\hat{V}$ is not a function of time, the Scrödinger equation can be simplified using the mathematical technique known as separation of variables. Let us present the wavefunction as the product of a spatial function and a time function:

$$\Psi(\mathbf{r}, \mathbf{R}, t) = \psi(\mathbf{r}, \mathbf{R})\tau(t). \tag{2.3}$$

Substituting these new functions into (2.1), two equations are obtained, one of which depends on the position of the particle independent of time and the other of which is a function of time alone. Let us consider the problems, when this separation is valid. The time-independent Scrödinger equation reads as:

$$\hat{H}\psi(\mathbf{r}, \mathbf{R}) = E\psi(\mathbf{r}, \mathbf{R}) \tag{2.4}$$

where E is the energy of the system.

The various solutions to (2.4) correspond to different stationary states of the molecular system. The one with the lowest energy is called the ground state. Equation (2.4) is a non-relativistic description of the system which is not valid when the velocities of particles approach the speed of light. Thus (2.4) does not give an accurate description of the core electrons in large nuclei.

The kinetic energy is defined as:

$$\hat{T} = -\frac{1}{2}\sum_k \frac{1}{m_k}\left(\frac{\partial^2}{\partial x_k^2} + \frac{\partial^2}{\partial y_k^2} + \frac{\partial^2}{\partial z_k^2}\right) = \frac{1}{2}\sum_k \frac{\hat{\mathbf{p}}_k^2}{m_k}, \tag{2.5}$$

where $\hat{\mathbf{p}}_k$ is the momentum operator of the particle k, and m_k is its mass.

The potential energy is defined by the Coulomb interaction between each pair of charged particles:

$$\hat{V} = \sum_{\substack{j < N \\ k < j}} \frac{e_j e_k}{|\mathbf{r}_j - \mathbf{r_k}|}, \tag{2.6}$$

where N is the number of the particles in the system, $|\mathbf{r}_j - \mathbf{r_k}|$ is the distance between the two particles j and k, and e_j and e_k are their charges. For an electron, the charge is -1, while for a nucleus, the charge is Z. Thus,

$$\hat{V} = -\sum_{\substack{i < N_e \\ I < N_n}} \frac{Z_I}{|\mathbf{r_i} - \mathbf{r_I}|} + \sum_{\substack{i < N_e \\ j < i}} \frac{1}{|\mathbf{r_i} - \mathbf{r_j}|} + \sum_{\substack{I < N_n \\ J < I}} \frac{Z_I Z_J}{|\mathbf{r_I} - \mathbf{r_J}|}, \tag{2.7}$$

where N_e is the number of electrons and N_n the number of nucleus in the system. The first term corresponds to electron–nucleus attraction, the second to electron–electron repulsion and the third to nucleus–nucleus repulsion.

2.3 The Born–Oppenheimer Approximation

If the nuclei move slowly with respect to the electrons, then it is possible to simplify the general molecular problem by separating nuclear and electronic motions.

This approximation is reasonable since the mass of a typical nucleus is thousands of times greater than that of an electron and the electrons react essentially instantaneously to changes in nuclear position. Thus, the electron distribution within a molecular system depends on the position of the nuclei, and not their velocities.

This approximation is called the Born–Oppenheimer approximation. The full Hamiltonian for the molecular system can be written as:

$$\hat{H} = \hat{T}^{\text{elec}}(\mathbf{r}) + \hat{T}^{\text{nucl}}(\mathbf{R}) + \hat{V}^{\text{nucl–elec}}(\mathbf{R}, \mathbf{r}) + \hat{V}^{\text{elec}}(\mathbf{r}) + \hat{V}^{\text{nucl}}(\mathbf{R}), \qquad (2.8)$$

where $\hat{T}^{\text{elec}}(\mathbf{r})$ is the electron kinetic energy, $\hat{T}^{\text{nucl}}(\mathbf{R})$ is the nucleon kinetic energy, $\hat{V}^{\text{nucl–elec}}(\mathbf{R}, \mathbf{r})$ is the nucleon–electron interaction, $\hat{V}^{\text{elec}}(\mathbf{r})$ describes the electron–electron interaction and $\hat{V}^{\text{nucl}}(\mathbf{R})$ is the nucleon–nucleon interaction. The Born–Oppenheimer approximation allows to separate the electronic and ionic subsystems, so one can construct an electronic Hamiltonian which neglects the kinetic energy term for the nuclei:

$$\hat{H}^{\text{elec}} = -\frac{1}{2} \sum_{i}^{N_e} \left(\frac{\partial^2}{\partial x_i^2} + \frac{\partial^2}{\partial y_i^2} + \frac{\partial^2}{\partial z_i^2} \right) - \sum_{\substack{i < N_e \\ I < N_n}} \left(\frac{Z_I}{|\mathbf{R_I} - \mathbf{r_i}|} \right)$$

$$+ \sum_{\substack{i < N_e \\ j < i}} \left(\frac{1}{|\mathbf{r_i} - \mathbf{r_j}|} \right) + \sum_{\substack{I < N_n \\ J < I}} \left(\frac{Z_I Z_J}{|\mathbf{R_I} - \mathbf{R_J}|} \right) \qquad (2.9)$$

This Hamiltonian describes the motion of electrons in the field of fixed nuclei:

$$\hat{H}^{\text{elec}} \psi^{\text{elec}}(\mathbf{r}, \mathbf{R}) = E^{\text{eff}}(\mathbf{R}) \psi^{\text{elec}}(\mathbf{r}, \mathbf{R}) \qquad (2.10)$$

The solution of (2.10) for the electronic wavefunction produces the effective nuclear potential function E^{eff}. It depends on the nuclear coordinates and describes the potential energy surface for the system.

Accordingly, E^{eff} is also used as the effective potential for the nuclear Hamiltonian:

$$\hat{H}^{\text{nucl}} = \hat{T}^{\text{nucl}}(\mathbf{R}) + E^{\text{eff}}(\mathbf{R}) \qquad (2.11)$$

This Hamiltonian is used in the Schrödinger equation for nuclear motion, describing the vibrational, rotational, and translational states of the ionic subsystem.

2.4 Properties of the Wavefunction

Let us focus entirely on the electronic problem. The superscript on all the operators and functions is further omitted.

As it is well known $|\psi|^2$ can be interpreted as the probability density for the particles it describes. Therefore, ψ has to be normalized. The integral of the probability over all space should be equal to the number of particles. Accordingly, ψ is multiplied by a constant such that:

$$\int_V |c\psi|^2 dV = n_{particles} \tag{2.12}$$

This is done because the Schrödinger equation is an eigenvalue equation, and in general, if f is a solution to an eigenvalue equation, than cf is also, for any value of c. For the Schrödinger equation, it is easy to show that $\hat{H}(c\psi) = c\hat{H}(\psi)$ and that $E(c\psi) = c(E\psi)$. Thus, if ψ is a solution to the Schrödinger equation, than $c\psi$ is as well.

Secondly, ψ must also be antisymmetric, meaning that it must change sign when two identical particles are interchanged. For a simple function, antisymmetry means that the following relation holds:

$$f(i, j) = -f(j, i) \tag{2.13}$$

For an electronic wavefunction, antisymmetry is a physical requirement following from the fact that electrons are fermions. More specifically, this requirement means that any valid wavefunction must satisfy the following condition:

$$\psi(\mathbf{r}_1, \ldots, \mathbf{r}_i, \ldots, \mathbf{r}_j, \ldots, \mathbf{r}_n) = -\psi(\mathbf{r}_1, \ldots, \mathbf{r}_j, \ldots, \mathbf{r}_i, \ldots, \mathbf{r}_n) \tag{2.14}$$

2.5 Hartree–Fock Theory

It is impossible to find an exact analytical solution of the Schrödinger equation for a multi-atomic system. However, a number of simplifying assumptions and procedures make its approximate solution possible.

At first, let us consider is the HF approximation [8]. The basic idea of this method is to replace the many-body problem by an effective one-body problem. Within the HF approximation the ground-state wavefunction is decomposed into a combination of the single-particle wavefunctions, which are often called molecular orbitals. Let us do so, and decompose the ground-state wave function of an N-body system of fermions, say electrons, into a combination of molecular orbitals: $\phi_1, \phi_2, \ldots$ To fulfill some of the conditions on ϕ discussed in previous section, a normalized, orthogonal set of molecular orbitals is chosen:

$$\int \phi_i^* \phi_i dV = 1 \tag{2.15}$$

$$\int \phi_i^* \phi_j dV = 0; \quad i \neq j \tag{2.16}$$

The simplest possible way of making ψ as a combination of these molecular orbitals is by forming their *Hartree product*:

$$\psi(\mathbf{r}) = \phi_1(\mathbf{r}_1)\phi_2(\mathbf{r}_1), \ldots, \phi_N(\mathbf{r_N}) \tag{2.17}$$

However, such a function is not antisymmetric, since interchanging two of the $\mathbf{r_i}$'s is equivalent to swapping the orbitals of two electrons, and does not result in a sign change. Hence, this Hartree product is an inadequate wavefunction.

The simplest antisymmetric function that is a combination of molecular orbitals is a determinant. Before forming it, however, one needs to account for a factor that was neglected so far: electron spin. Electrons can have spin up $(+\frac{1}{2})$ or down $(-\frac{1}{2})$. Equation (2.17) assumes that each molecular orbital holds only one electron. However, most calculations are closed shell calculations, using doubly occupied orbitals, holding two electrons of opposite spin. For the moment, let us limit the discussion to this case.

Two spin functions, α and β, are defined as follows:

$$\begin{aligned}
\alpha(\uparrow) = 1 \quad \alpha(\downarrow) = 0 \\
\beta(\uparrow) = 0 \quad \beta(\downarrow) = 1
\end{aligned} \tag{2.18}$$

The α function is 1 for a spin up electron, and the β function is 1 when the electron is spin down. The notations $\alpha(i)$ and $\beta(i)$ designate the values of α and β for electron i.

Multiplying a molecular orbital function by α or β includes electron spin as part of the overall electronic wavefunction ψ. The product of the molecular orbital and a spin function is defined as a *spin orbital*, a function of both the electron's location and its spin. Note that these spin orbitals are also orthonormal when the component molecular orbitals are.

A closed shell wavefunction can be build now by defining $N/2$ molecular orbitals for a system with N electrons, and then assigning electrons to these orbitals in pairs of opposite spin:

$$\psi(\mathbf{r}) = \frac{1}{\sqrt{N!}}
\begin{vmatrix}
\phi_1(\mathbf{r}_1)\alpha(1) & \phi_1(\mathbf{r}_1)\beta(1) & \ldots & \phi_{\frac{N}{2}}(\mathbf{r}_1)\alpha(1) & \phi_{\frac{N}{2}}(\mathbf{r}_1)\beta(1) \\
\phi_1(\mathbf{r}_2)\alpha(2) & \phi_1(\mathbf{r}_2)\beta(2) & \ldots & \phi_{\frac{N}{2}}(\mathbf{r}_2)\alpha(2) & \phi_{\frac{N}{2}}(\mathbf{r}_2)\beta(2) \\
& & \vdots & & \vdots \\
\phi_1(\mathbf{r}_i)\alpha(i) & \phi_1(\mathbf{r}_i)\beta(i) & \ldots & \phi_{\frac{N}{2}}(\mathbf{r}_i)\alpha(i) & \phi_{\frac{N}{2}}(\mathbf{r}_i)\beta(i) \\
\phi_1(\mathbf{r}_j)\alpha(j) & \phi_1(\mathbf{r}_j)\beta(j) & \ldots & \phi_{\frac{N}{2}}(\mathbf{r}_j)\alpha(j) & \phi_{\frac{N}{2}}(\mathbf{r}_j)\beta(j) \\
& & \vdots & & \vdots \\
\phi_1(\mathbf{r_N})\alpha(N) & \phi_1(\mathbf{r_N})\beta(N) & \ldots & \phi_{\frac{N}{2}}(\mathbf{r_N})\alpha(N) & \phi_{\frac{N}{2}}(\mathbf{r_N})\beta(N)
\end{vmatrix} \tag{2.19}$$

The determinant (2.19) is also called the Slater determinant. Each row is formed by representing all possible assignments of electron i to all orbital–spin combinations.

The initial factor is necessary for normalization. Swapping two electrons corresponds to interchanging two rows of the determinant, which has the effect of changing its sign. Note, that the wavefunction (2.19) also accounts for the Pauli principle, which says that two or more fermions can not be found in the same quantum state. Two or more identical electrons correspond to two or more identical rows in the Slater determinant, what makes it equal to zero. Further a notation $\psi(\mathbf{r}) = |a, b, \ldots, n\rangle$ is used.

Let us introduce another notation $\phi_i(\mathbf{r}_j, s_j) \equiv \phi_i(j)$—the molecular orbital of the i-th electron with spin. Here i and j run over all integer values from 1 to N. With this new notation one obtains: $\phi_i(\mathbf{r}_j, s_j) = \phi_{\frac{i+1}{2}}(\mathbf{r}_j)\alpha(j)$ for spin up, and $\phi_i(\mathbf{r}_j, s_j) = \phi_{\frac{i}{2}}(\mathbf{r}_j)\beta(j)$ for spin down.

In order to calculate the energy levels of the system with N electrons, it is necessary to evaluate matrix elements of the Hamiltonian between the antisymmetric states. The Hamiltonian of the system with N electrons reads as:

$$\hat{H} = -\frac{1}{2}\sum_{i=1}^{N}\nabla_i^2 - \sum_{i=1}^{N}\frac{Z}{r_i} + \sum_{i<j}^{N}\frac{1}{r_{ij}} \tag{2.20}$$

Here the first term represents the kinetic energy of the electrons, the second term represents their attraction to the ionic core, and the last term represents the interelectron interaction. The Hamiltonian (2.20) includes one-electron operators of the type Z/r_i, which act on the coordinates of one electron, and two-electron operators of the kind $1/r_{ij}$. Therefore the matrix elements of one- and two-electron operators between determinants of orthonormal functions are needed.

Let us consider first a general single-electron operator, which may be written:

$$F = \sum_{i=1}^{N} f(i) \tag{2.21}$$

where $f(i)$ acts only on the coordinates of the i-th electron. For simplicity, the consideration is restricted to a two electron system for which

$$F = f(1) + f(2) \tag{2.22}$$

The diagonal matrix elements of F for the antisymmetric wavefunction $|ab\rangle$ are:

$$\langle ab|F|ab\rangle = \frac{1}{2}\int\int [\phi_a(1)\phi_b(2) - \phi_a(2)\phi_b(1)]^*$$
$$\times [f(1) + f(2)][\phi_a(1)\phi_b(2) - \phi_a(2)\phi_b(1)]\mathrm{d}\mathbf{r}_1\mathrm{d}\mathbf{r}_2, \tag{2.23}$$

where $\mathrm{d}\mathbf{r}_1$ and $\mathrm{d}\mathbf{r}_2$, denote the volume elements, and their integrations also represent summations over the spin coordinates, Cross terms of the kind

$$\int\int \phi_a^*(1)\phi_b^*(2)f(1)\phi_a(2)\phi_b(1)\mathrm{d}\mathbf{r}_1\mathrm{d}\mathbf{r}_2 \tag{2.24}$$

are obviously zero, since $f(1)$ operates only on the first wavefunction and $\phi_b(2)$ and $\phi_a(2)$ are assumed to be orthogonal. Furthermore, by interchanging the coordinates of the first and the second electron, one may easily see that

$$\int\int \phi_a^*(1)\phi_b^*(2)f(1)\phi_a(1)\phi_b(2)\mathrm{dr_1dr_2} = \int\int \phi_a^*(1)\phi_b^*(2)f(2)\phi_a(2)\phi_b(1)\mathrm{dr_1dr_2}$$

$$(2.25)$$

So (2.23) reduces to

$$\langle ab|F|ab\rangle = \int\int \phi_a^*(1)\phi_b^*(2)[f(1)+f(2)]\phi_a(1)\phi_b(2)\mathrm{dr_1dr_2}$$
$$= \langle a|f|a\rangle + \langle b|f|b\rangle. \tag{2.26}$$

The nondiagonal matrix element between two determinantal states, which differ by a single state, can be shown in a similar way to be

$$\langle ab|F|ac\rangle = \langle b|f|c\rangle, \tag{2.27}$$

and finally, if both are different, one gets

$$\langle ab|F|cd\rangle = 0, \tag{2.28}$$

A two-electron operator can be written generally

$$G = \sum_{i<j} g(i,j) \tag{2.29}$$

where $g(i,j)$ acts on the i-th and the j-th electrons, and the summation includes each pair of electrons. For a two-electron system this is simply $G = g(1,2)$. The diagonal matrix element of G is then

$$\langle ab|G|ab\rangle = \frac{1}{2}\int\int [\phi_a(1)\phi_b(2) - \phi_a(2)\phi_b(1)]^* g(1,2) \tag{2.30}$$
$$\times [\phi_a(1)\phi_b(2) - \phi_a(2)\phi_b(1)]\mathrm{dr_1dr_2}$$
$$= \frac{1}{2}\int\int [\phi_a^*(1)\phi_b^*(2)g(1,2)\phi_a(1)\phi_b(2)$$
$$- \phi_a^*(1)\phi_b^*(2)g(1,2)\phi_a(2)\phi_b(1)$$
$$- \phi_a^*(2)\phi_b^*(1)g(1,2)\phi_a(1)\phi_b(2)$$
$$+ \phi_a^*(2)\phi_b^*(1)g(1,2)\phi_a(2)\phi_b(1)]\mathrm{dr_1dr_2}$$

Since the two-electron interaction $g(1,2)$ is symmetric with respect to an interchange of the coordinates of the two electrons $(1 \leftrightarrow 2)$, the first and the fourth terms in this expansion are equal, and similarly the second and the third terms are equal. So the matrix element may be written simply

$$\langle ab|G|ab\rangle = \langle ab|g|ab\rangle - \langle ba|g|ab\rangle. \tag{2.31}$$

The symbols on the right represent here matrix elements with ordinary product functions. The first matrix element is called the *direct* term and the second matrix element the *exchange* term. The exchange matrix element would not occur if one uses product functions $\phi_a(1)\phi_b(2)$ rather than the proper antisymmetric wavefunctions.

The results obtained above may be generalized to N-electron systems. For this purpose a special notation is used. Let us allow Greek letters to stand for ordered sets of quantum numbers representing Slater determinants. So for instance, let α correspond to the quantum numbers $a, b, \ldots, n$. Then, the determinantal state $|ab, \ldots, n\rangle$ can be written simply $|\alpha\rangle$. Single-particle functions, which appear in the determinant, are called *occupied* orbitals and the remaining functions of the set are called *excited* or *virtual* orbitals. The notation $|\alpha_a^r\rangle$ will be used to denote a determinant for which an occupied orbital a in α is replaced by the virtual orbital r. Similarly, double substitutions for which two electrons (here a and b) are excited from the sea of occupied orbitals can be written $|\alpha_{ab}^{rs}\rangle$.

Using this notation the formulas for the matrix elements of one- and two-particle operators between determinantal states of a many-particle system can be generalized in the following way.

For diagonal elements:

$$\langle\alpha|F|\alpha\rangle = \sum_a^{\text{occ}} \langle a|f|a\rangle \tag{2.32}$$

$$\langle\alpha|G|\alpha\rangle = \sum_{a<b}^{\text{occ}} (\langle ab|g|ab\rangle - \langle ba|g|ab\rangle) \tag{2.33}$$

where the sums run over orbitals a and b that are occupied in $|\alpha\rangle$.

For elements between states which differ by the quantum numbers of a single orbital

$$\langle\alpha_a^r|F|\alpha\rangle = \langle r|f|a\rangle \tag{2.34}$$

$$\langle\alpha_a^r|G|\alpha\rangle = \sum_b^{\text{occ}} (\langle rb|g|ab\rangle - \langle br|g|ab\rangle) \tag{2.35}$$

For elements between states which differ by the quantum numbers of two orbitals:

$$\langle\alpha_{ab}^{rs}|F|\alpha\rangle = 0 \tag{2.36}$$

$$\langle\alpha_{ab}^{rs}|G|\alpha\rangle = \langle rs|g|ab\rangle - \langle sr|g|ab\rangle \tag{2.37}$$

All matrix elements of F and G between states for which more than two quantum numbers are different vanish.

Equations (2.32)–(2.37) may be used to evaluate the matrix elements of the atomic Hamiltonian (2.20). The expectation value of the total energy for a state represented by a Slater determinant $|\alpha\rangle$ is

$$\langle E\rangle = \langle\alpha|H|\alpha\rangle = \left\langle\alpha\left|\sum_{i=1}^{N}\left(-\frac{1}{2}\nabla_i^2 - \frac{Z}{r_i}\right) + \sum_{i<j}^{N}\frac{1}{r_{ij}}\right|\alpha\right\rangle. \tag{2.38}$$

According to the variational principle, the "best" determinant for the ground state can be determined by minimizing this expectation value. A necessary condition is then that expectation value be stationary with respect to small changes in the form of the occupied orbitals, and this condition is used to derive the HF equations as follows:

Small changes in the occupied orbitals (a) can be expressed by means of small admixtures of virtual orbitals (r)

$$|a\rangle \to |a\rangle + \eta|r\rangle \tag{2.39}$$

where η is a small, real number. This leads to an admixture of single substitutions $|\alpha_a^r\rangle$ into $|\alpha\rangle$

$$|a\rangle \to |a\rangle + \eta|\alpha_a^r\rangle \tag{2.40}$$

and a corresponding change in the expectation value of the energy

$$\langle E\rangle \to \langle E\rangle + \eta\left(\langle\alpha_a^r|H|\alpha\rangle + \langle\alpha|H|\alpha_a^r\rangle\right), \tag{2.41}$$

neglecting terms quadratic in η. Since H is a hermitian operator, the two matrix elements above are complex conjugates of each other. With the conventions used here, the elements are real and hence equal. The energy is stationary if

$$\langle\alpha_a^r|H|\alpha\rangle = 0. \tag{2.42}$$

This condition is called *Brillouin's* theorem, implies that the Hamiltonian H has no matrix elements between $|\alpha\rangle$ and states obtained from $|\alpha\rangle$ by a single substitution.

Using (2.34) and (2.35) the HF condition (2.42) may be written out explicitly in terms of one- and two-particle matrix elements,

$$\left\langle r\left|-\frac{1}{2}\nabla^2 - \frac{Z}{r}\right|a\right\rangle + \sum_{b}^{\mathrm{occ}}\left(\left\langle rb\left|\frac{1}{r_{ij}}\right|ab\right\rangle - \left\langle br\left|\frac{1}{r_{ij}}\right|ab\right\rangle\right) = 0 \tag{2.43}$$

In order to write (2.43) in a more simple form, let us define a HF operator (H_{HF}) and potential (U_{HF}) by the equations

$$H_{\mathrm{HF}} = -\frac{1}{2}\nabla^2 - \frac{Z}{r} + U_{\mathrm{HF}} \tag{2.44}$$

$$\langle j|U_{\mathrm{HF}}|j\rangle = \sum_{b}^{\mathrm{occ}} \left(\left\langle rb \left| \frac{1}{r_{ij}} \right| ab \right\rangle - \left\langle br \left| \frac{1}{r_{ij}} \right| ab \right\rangle \right) \tag{2.45}$$

where the sum b runs over all of the orbitals occupied in the determinant $|\alpha\rangle$. Then the condition (2.43) to be satisfied becomes simply

$$\langle r|H_{\mathrm{HF}}|a\rangle = 0 \tag{2.46}$$

where a is an occupied and r a virtual orbital. Using the completeness relation $\left(\sum_i |i\rangle\langle i| = 1 \right)$, this leads to the equation

$$H_{\mathrm{HF}}|a\rangle = \sum_{i} |i\rangle\langle i|H_{\mathrm{HF}}|a\rangle = \sum_{b}^{\mathrm{occ}} |b\rangle\langle b|H_{\mathrm{HF}}|a\rangle, \tag{2.47}$$

where i runs over all orbitals and b over occupied ones. Thus, when acting on an occupied orbital the HF operator produces only occupied orbitals. It follows directly from the symmetry of th Coulomb interaction that $\langle a|H_{\mathrm{HF}}|b\rangle = \langle b|H_{\mathrm{HF}}|a\rangle$, which means that the HF operator is hermitian. Furthermore, it can be shown that this operator is invariant for unitary transformation. Therefore, a set of orbitals, where H_{HF} is diagonal, can be found:

$$H_{\mathrm{HF}}|a'\rangle = \varepsilon_a'|a'\rangle. \tag{2.48}$$

This is the normal form of the general HF equation. Using (2.44) the HF equation (2.48) can be written out explicitly

$$\left(-\frac{1}{2}\nabla^2 - \frac{Z}{r} + U_{\mathrm{HF}} \right) |a\rangle = \varepsilon_a |a\rangle. \tag{2.49}$$

Each term here can be given a simple physical interpretation. The first term represents the kinetic energy of electron a and Z/r its attraction to the nucleus. The potential U_{HF} represents the average Coulomb and the exchange interaction of electron a with other electrons in the atom.

For an effective numerical solution of (2.49) and similar equations, the molecular orbitals, ϕ_i, are often approximated by a linear combination of a pre-defined set of single-electron functions, χ_μ, known as basis functions. This expansion reads as follows:

$$\phi_i = \sum_{\mu=1}^{N} c_{\mu i} \chi_\mu, \tag{2.50}$$

where coefficients $c_{\mu i}$ are the molecular orbital expansion coefficients, N is the number of basis functions, which are chosen to be normalized.

The basis functions χ_μ are defined as linear combinations of primitive gaussians:

$$\chi_\mu = \sum_p d_{\mu p} g_p, \tag{2.51}$$

where $d_{\mu p}$ are fixed constants within a given basis set, the primitive gaussians, $g_p = g(\alpha, \mathbf{r})$, are the gaussian-type atomic functions having the following form:

$$g(\alpha, \mathbf{r}) = c x^n y^m z^l e^{-\alpha r^2} \tag{2.52}$$

Here, c is the normalization constant. The choice of the integers n, m and l defines the type of the primitive gaussian function: s, p, d or f (for details see [9]).

Here are three representative gaussian functions (s, p_y and d_{xy} types, respectively):

$$g_s(\alpha, \mathbf{r}) = \left(\frac{2\alpha}{\pi}\right)^{3/4} e^{-\alpha r^2} \tag{2.53}$$

$$g_y(\alpha, \mathbf{r}) = \left(\frac{128\alpha^5}{\pi^3}\right)^{1/4} y e^{-\alpha r^2} \tag{2.54}$$

$$g_{xy}(\alpha, \mathbf{r}) = \left(\frac{2048\alpha^7}{\pi^3}\right)^{1/4} xy e^{-\alpha r^2} \tag{2.55}$$

In the calculations presented in the next chapter of the thesis the standard 6-31G (d), 6-31G (2d, p) and 6-31++G(d, p) basis sets were used. The detailed information on the basis constants can be found for example in [9, http://www.emsl.pnl.gov/forms/basisform.html].

All of these constructions result in the following expansion for molecular orbitals:

$$\phi_i = \sum_{\mu=1}^{N} c_{\mu i} \chi_\mu = \sum_{\mu=1}^{N} c_{\mu i} \left(\sum_p d_{\mu p} g_p\right) \tag{2.56}$$

The problem has now become how to solve for the set of molecular orbital expansion coefficients, $c_{\mu i}$. HF theory takes advantage of the variational principle, which says that for the ground state of any antisymmetric normalized function of the electronic coordinates, which is denoted by Ξ, the expectation value for the energy corresponding to Ξ will always be greater than the energy for the exact wavefunction:

$$E(\Xi) > E(\psi); \quad \Xi \neq \psi \tag{2.57}$$

In other words, the energy of the exact wavefunction serves as a lower bound to the energies calculated by any other normalized antisymmetric function. Thus the problem becomes one of finding the set of coefficients that minimize the energy of the resultant wavefunction.

The variational principle leads to the following equations describing the molecular orbital expansion coefficients, c_{vi}, known also as the Roothaan and Hall equations:

$$\sum_{v=1}^{N}(H_{\mu v} - \varepsilon_i S_{\mu v})c_{vi} = 0 \quad \mu = 1, 2, \ldots, N \tag{2.58}$$

Being written in the matrix form, this equation reads as:

$$HC = SC\varepsilon, \tag{2.59}$$

where each element is a matrix. Here, ε is a diagonal matrix of orbital energies, each of its elements ε_i is the single-electron energy of the molecular orbital ψ_i, H is the Hamiltonian matrix as it follows from (2.48), S is the overlap matrix, describing the overlap between orbitals. For more details regarding this formalism see [9].

Equation (2.59) is none linear and must be solved iteratively. The procedure which does so is called the *self-consistent field* (SCF) method.

The above written equations consider the restricted HF method. For the open shell systems, the unrestricted HF method has to be used. In this case, the alpha and beta electrons with spins up and down are assigned to different orbitals, resulting in two sets of molecular orbital expansion coefficients:

$$\begin{aligned}
\phi_i^\alpha &= \sum_{\mu=1}^{N} c_{\mu i}^\alpha \chi_\mu \\
\phi_i^\beta &= \sum_{\mu=1}^{N} c_{\mu i}^\beta \chi_\mu,
\end{aligned} \tag{2.60}$$

The two sets of coefficients result in two sets of the Hamiltonian matrices and the two sets of orbitals.

2.6 Density Functional Theory

DFT is another approach, which accounts for many-electron correlation interaction. It is based upon a strategy of modelling electron correlation via general fundamental functionals of the electron density.

Within the DFT one has to solve the Kohn–Sham equations, which read as (see e.g. [10–16])

$$\left(\frac{\hat{p}^2}{2} + U_{\text{ions}} + V_H + V_{xc}\right)\psi_i = \varepsilon_i\psi_i, \tag{2.61}$$

where the first term represents the kinetic energy of the i-th electron, and U_{ions} describes its attraction to the ions in the cluster, V_H is the Hartree part of the inter-electronic interaction:

$$V_H(\mathbf{r}) = \int \frac{\rho(\mathbf{r}')}{|\mathbf{r} - \mathbf{r}'|} d\mathbf{r}' , \qquad (2.62)$$

where $\rho(\mathbf{r})$ is the electron density:

$$\rho(\mathbf{r}) = \sum_{\nu=1}^{N_e} |\psi_i(\mathbf{r})|^2 , \qquad (2.63)$$

and V_{xc} is the local exchange-correlation potential, ψ_i are the electronic orbitals and N_e is the number of electrons in the cluster.

The exchange-correlation potential is defined as the functional derivative of the exchange-correlation energy functional:

$$V_{xc} = \frac{\delta E_{xc}[\rho]}{\delta\rho(\mathbf{r})}, \qquad (2.64)$$

One of the well-known approximations is the Gunnarsson and Lundqvist model [17]. It is based upon the calculation of the self-energy of an electron for the homogeneous electron gas. The local Gunnarsson and Lundqvist exchange-correlation energy density functional reads as:

$$E_{xc}^{GL} = -\frac{3}{4}\left(\frac{9}{4\pi^2}\right)^{1/3}\frac{1}{r_s(\mathbf{r})} - 0.0333\, G\left(\frac{r_s(\mathbf{r})}{11.4}\right). \qquad (2.65)$$

Here $r_s(\mathbf{r}) = (3/4\pi\rho_{\text{el}}(\mathbf{r}))^{1/3}$ is a local Wigner–Seitz radius, while $\rho_{\text{el}}(\mathbf{r})$ is the electron density in the cluster, and the function $G(x)$ is defined by following relation:

$$G(x) = (1 + x^3)\ln\left(1 + \frac{1}{x}\right) - x^2 + \frac{x}{2} - \frac{1}{3}. \qquad (2.66)$$

The first and the second terms in (2.65) corresponds to the exchange and correlation interaction respectively. The exchange-correlation energy density E_{xc}^{GL}, defines the LDA exchange-correlation potential V_{xc}^{GL} as

$$V_{xc}^{GL} = \frac{\delta\left[\rho_{\text{el}}(\mathbf{r})E_{xc}^{GL}(\rho_{\text{el}}(\mathbf{r}))\right]}{\delta\rho_{\text{el}}(\mathbf{r})} =$$
$$-\left(\frac{9}{4\pi^2}\right)^{1/3}\frac{1}{r_s(\mathbf{r})} - 0.0333\,\ln\left(1 + \frac{11.4}{r_s(\mathbf{r})}\right). \qquad (2.67)$$

The approximate functionals employed by DFT methods often separate the exchange-correlation energy into two parts, referred to as *exchange* and *correlation* parts:

$$E_{xc}[\rho] = E_x(\rho) + E_c(\rho) \tag{2.68}$$

Both parts are the functionals of the electron density, which can be of two distinct types: either *local* functional depending on only the electron density ρ or *gradient-corrected* functionals depending on both ρ and its gradient, $\nabla\rho$.

In literature, there is a variety of exchange correlation functionals. Below, are presented only those, which are related to the calculation performed in the thesis.

The local exchange functional is virtually always defined as follows:

$$E_x^{LDA} = -\frac{3}{2}\left(\frac{3}{4\pi}\right)^{1/3}\int \rho^{4/3}d^3\mathbf{r}. \tag{2.69}$$

This form was developed to reproduce the exchange energy of a uniform electron gas. By itself, however, it is not sufficient for the adequate description of atomic clusters.

The gradient-corrected exchange functional introduced by Becke [18] and based on the LDA exchange functional reads as:

$$E_x^{B88} = E_x^{LDA} - \gamma\int \frac{\rho^{4/3}x^2}{1 + 6\gamma\,Sinh^{-1}x}d^3\mathbf{r}, \tag{2.70}$$

where $x = \rho^{-4/3}|\nabla\rho|$ and $\gamma = 0.0042$ is a parameter chosen to fit the known exchange energies of the noble gas atoms.

Analogously to the above written gradient-corrected exchange functionals, there are gradient-corrected correlation functionals. For example, here is the correlation functional introduced by Burke et al. [19]:

$$E_c^{PW91} = \int \rho\epsilon_c\,(r_s(\rho(\mathbf{r})), \zeta)\,d^3\mathbf{r}$$

$$r_s = \left[\frac{3}{4\pi\rho}\right]^{1/3}$$

$$\zeta = \frac{\rho_\alpha - \rho_\beta}{\rho_\alpha + \rho_\beta}$$

$$\epsilon_c(r_s, \zeta) = \epsilon_c(\rho, 0) + a_c(r_s)\frac{f(\zeta)}{f''(0)}(1 - \zeta^4) + [\epsilon_c(\rho, 1) - \epsilon_c(\rho, 0)]\,f(\zeta)\zeta^4$$

$$f(\zeta) = \frac{(1 + \zeta)^{4/3} + (1 - \zeta)^{4/3} - 2}{2^{4/3} - 2}, \tag{2.71}$$

where ρ_α is used to refer to the alpha spin density, ρ_β to refer to the beta spin density, ρ to refer to the total electron density, $(\rho_\alpha + \rho_\beta)$. r_s is the local Wigner–Seitz radius. ζ is the relative spin polarization. $\zeta = 0$ corresponds to equal α and β densities, $\zeta = 1$ corresponds to all α density and $\zeta = -1$ corresponds to all β density.

In the pure DFT, an exchange functional usually pairs with a correlation functional. For example, the well-known BLYP functional pairs Becke's gradient-corrected exchange functional (2.70) with the gradient-corrected correlation functional of Lee et al. [20].

The gradient-corrected correlation functional of Lee, Yang and Parr reads as:

$$
\begin{aligned}
E_c^{\mathrm{LYP}} = -\,a \int \frac{\gamma(\mathbf{r})}{1 + d\rho^{-1/3}} & \left\{ \rho + 2b\rho^{-5/3} \left[2^{2/3} C_F \rho_\alpha^{8/3} \right. \right. \\
& + 2^{2/3} C_F \rho_\beta^{8/3} - \rho t_W + \frac{1}{9}\left(\rho_\alpha t_W^\alpha + \rho_\beta t_W^\beta \right) \\
& \left. \left. + \frac{1}{18}\left(\rho_\alpha \nabla^2 \rho_\alpha + \rho_\beta \nabla^2 \rho_\beta \right) \right] e^{-c\rho^{-1/3}} \right\} d^3\mathbf{r}
\end{aligned}
\tag{2.72}
$$

where

$$
\gamma(\mathbf{r}) = 2\left(1 - \frac{\rho_\alpha^2(\mathbf{r}) + \rho_\beta^2(\mathbf{r})}{\rho^2(\mathbf{r})} \right)
$$

$$
t_W(\mathbf{r}) = \frac{1}{8}\frac{|\nabla \rho(\mathbf{r})|^2}{\rho((r))} - \frac{1}{8}\nabla^2 \rho
$$

$$
C_F = \frac{3}{10}\left(3\pi^2 \right)^{2/3}
\tag{2.73}
$$

$t_W(\mathbf{r})$ is the local kinetic-energy density, $t_W^\alpha(\mathbf{r})$ and $t_W^\beta(\mathbf{r})$ are the kinetic-energy densities of the α-spin and β-spin electron densities, respectively. The parameters in equation (2.73) are as follows: $a = 0.04918$, $b = 0.132$, $c = 0.2533$ and $d = 0.349$.

In spite of the success of the pure DFT theory in many cases, one has to admit that the HF theory accounts for the electron exchange the most naturally and precisely. Thus, Becke has suggested [18] functionals which include a mixture of HF and DFT exchange along with DFT correlations, conceptually defining E_{xc} as:

$$
E_{xc}^{\mathrm{mix}} = c_{\mathrm{HF}} E_x^{\mathrm{HF}} + c_{\mathrm{DFT}} E_{xc}^{\mathrm{DFT}},
\tag{2.74}
$$

where c_{HF} and c_{DFT} are constants. Following this idea, a Becke-type three parameter functional (B3LYP) can be defined as follows:

$$
\begin{aligned}
E_{xc}^{\mathrm{B3LYP}} = {} & E_x^{\mathrm{LDA}} + c_0(E_x^{\mathrm{HF}} - E_X^{\mathrm{LDA}}) + c_x(E_x^{\mathrm{B88}} - E_x^{\mathrm{LDA}}) \\
& + E_c^{\mathrm{VWN3}} + c_c(E_c^{\mathrm{LYP}} - E_c^{\mathrm{VWN3}})
\end{aligned}
\tag{2.75}
$$

Here, $c_0 = 0.2$, $c_x = 0.72$ and $c_c = 0.81$ are constants, which were defined by fitting to the atomization energies, ionization potentials, proton affinities and first-row atomic energies [9]. E_x^{LDA} and E_x^{B88} are defined in (2.69) and (2.70), respectively. E_x^{HF} is the functional corresponding to HF equations (2.48). E_c^{VWN3} is the so-called Vosko–Wilk–Nusair (VWN) functional [21], which reads as:

Table 2.1 Constants for the Vosko–Wilk–Nusair parameterization

Parameter	I	II	III
A_i	0.06218	0.03109	−0.033774
b_i	3.72744	7.06042	1.131071
c_i	12.93520	18.05780	13.004500
x_{0i}	−0.10498	−0.32500	−0.004758

$$E_c^{\text{VWN3}} = \int \rho \epsilon_c^{\text{VWN3}}(\rho_\alpha, \rho_\beta) d^3\mathbf{r}$$

$$\epsilon_c^{\text{VWN3}}(\rho_\alpha, \rho_\beta) = \epsilon_I(\rho_\alpha, \rho_\beta) + \Delta\epsilon_c(r_s, \xi)$$

$$\epsilon_i = A_i \left(Ln \frac{r_s}{X_i(\sqrt{r_s})} + \frac{2b_i}{Q_i} Tan^{-1} \left(\frac{Q_i}{2\sqrt{r_s} + b_i} \right) \right.$$

$$- \frac{b_i x_{0i}}{X_i(x_{0i})} \left(Ln \frac{(\sqrt{r_s} - x_{0i})^2}{X(\sqrt{r_s})} \right.$$

$$\left. \left. + \frac{2(b_i + 2x_{0i})}{Q_i} Tan^{-1} \frac{Q_i}{2\sqrt{r_s} + b_i} \right) \right)$$

$$\Delta\epsilon_c(r_s, \xi) = \epsilon_{III}(\rho_\alpha, \rho_\beta) \frac{f(\zeta)}{f''(0)} (1 + \beta(r_s)\zeta^4) \tag{2.76}$$

$$\beta(r_s) = \frac{f''(0)}{\epsilon_{III}(\rho_\alpha, \rho_\beta)} \Delta\epsilon(r_s, 1) - 1$$

$$\Delta\epsilon_c(r_s, 1) = \epsilon_I(\rho_\alpha, \rho_\beta) - \epsilon_{II}(\rho_\alpha, \rho_\beta)$$

$$Q_i = \sqrt{4c_i - b_i^2}$$

$$X(x) = x^2 + b_i x + c_i$$

$$r_s = \left[\frac{3}{4\pi\rho} \right]^{1/3},$$

where the constants A_i, b_i, c_i and x_{0i} are given in the Table 2.1. The gradient-corrected correlation functional of Lee, Yang and Parr, E_c^{LYP} is defined in (2.73). Note that instead of E_c^{VWN3} and E_c^{LYP} in (2.75) one can also use the Perdew and Wang correlation functional (2.72) to obtain the so-called B3PW91 functional, which is also used for the calculation.

Post HF perturbation theories and the density functional approximation are the two different theoretical schemes for the solution of many-electron correlation problem based on different physical principles. The important feature of the density functional method consists in the fact, that this method takes into account many-electron correlations via the phenomenological exchange-correlation potential. However, so far, there has not been found the unique potential, universally applicable for different systems and conditions. As a result there is a "zoo of potentials" (see e.g. D. Salahub, session LXXIII, in [15]) valid for special cases. These potentials of course do exist in principle as unique quantities but are not actually understood, so alone they can not serve as a satisfactory basis for achieving a physical interpretation.

2.7 Molecular Mechanics Approach: a Way to Overcome the Complexity of Quantum Mechanics

It is not possible to apply straightforwardly the methods of quantum mechanics for the description of the dynamical behavior of large molecular systems such as proteins due to the fact that ab initio methods are computationally demanding. However, one can distinguish the principal coordinates in the molecules that correspond to the quantum nature of the covalent chemical bonds in the system. These coordinates are usually the distances between atoms, the angles between two neighboring chemical bonds and the dihedral angles that correspond to the twisting along chemical bonds. The dynamics of the system in the coordinates of bond lengths, angles between bonds and dihedral angles can be described classically at moderate temperatures ($\hbar\omega \ll 2kT$, i.e. at temperatures at which one can omit the quantum corrections to the vibrations). Such a description implies the construction of a classical Hamiltonian of the system that describes the interactions between the atoms. The classical Hamiltonian for the description of the dynamics of atoms in a molecule is usually constructed on the basis of so call potential of Molecular Mechanics (which is discussed in Sect. 4.2). The parameters of the Molecular Mechanics potential that describe the "stiffness" of chemical bonds, angles between bonds, etc. are usually obtained on the basis of quantum mechanical calculations of the fragments of a large molecule. For instance, the substantial part of the parameters of a widely used CHARMM forcefield [22] are obtained from calculations of alanine dipeptide using a HF theory with a 6-31G(d) basis set. HF theory and various basis sets were introduced in Sect. 2.5. The discussion of the Molecular Mechanics potential and it accuracy are presented in Sects. 4.2 and 5.3 of the thesis correspondingly.

References

1. Russell, D., Lasker, K., Phillips, J., Schneidman-Duhovny, D., Velaszquez-Muriel, J., & Sali, A. (2009). The structural dynamics of macromolecular processes. *Current Opinion in Cell Biology, 21*, 1–12.
2. Casella, M., & Peraro, M. (2008). Challenges and perspectives in biomolecular simulations: from the atomistic picture to multiscale modeling. *Current Opinion in Cell Biology, 18*, 630–640.
3. Solov'yov, A., Surdutovich, E., Scifoni, E., Mishustin, I., & Greiner, W. (2009). Physics of ion beam cancer therapy: a multi-scale approach. *Physical Review E, 011909*, 1–7.
4. Marx, D., & Hutter, J. (2000). *Structure and mechanism in protein science: a guide to enzyme catalysis and protein folding.* Julich: John von Neumann Institute for Computing.
5. Knowles, P., Schutz, M., & Werner, H.-J. (2000). Ab initio methods for electron correlation in molecules. In J. Grotendorst (Ed.), *Modern methods and algorithms of quantum chemistry* (pp. 397–434). Julich, Germany: John von Neumann Institute for Computing.
6. Marques, M., & Gross, E. (2004). Time-dependent density functional theory. *Annual Review of Physical Chemistry, 55*, 427–455.
7. Landau, L., & Lifshitz, E. (1959). *Quantum mechanics.* London-Paris: Pergamon Press.
8. Lindgren, I., & Morrison, J. (1982). *Atomic many-body theory.* Berlin: Springer.

9. Foresman, J. B., & leen Frisch, A. (1996). *Exploring chemistry with electronic structure methods*. Pittsburgh, PA: Gaussian Inc.
10. de Heer, W. (1993). The physics of simple metal-clusters—experimental aspects and simple-models. *Review of Modern Physics, 65*, 611.
11. Brack, M. (1993). The physics of simple metal-clusters—self-consistent jellium model and semiclassical approaches. *Review of Modern Physics, 65*, 677–732.
12. Haberland, H. (Ed.). (1994). Clusters of atoms and molecules, theory, experiment and clusters of atoms. In *Springer Series in Chemical Physics*. Berlin: Springer.
13. Näher, U., Björnholm, S., Frauendorf, S., Garcias, F., & Guet, C. (1997). Fission of metal clusters. *Physics Reports, 285*, 245–320.
14. Ekardt, W. (Ed.) (1999). *Metal clusters*. New York: Wiley.
15. Clementi, E., & Corongiu G. (Eds.) (2001). *Atomic clusters and nanoparticles. NATO Advanced Study Institute, les Houches Session LXXIII, les Houches, 2000*. Berlin: EDP Sciences and Springer Verlag.
16. Fiolhais, C., Nogueira, F., & Marques, M. (Eds.). (2003). A primer in density functional theory. In *Springer lecture notes in physics*. Berlin: Springer.
17. Gunnarsson, O., & Lundqvist, B. (1976). Exchange and correlation in atoms, molecules, and solids by spin-density functional formalism. *Physical Review B, 13*, 4274–4298.
18. Becke, A. (1988). Density-functional exchange-energy approximation with correct asymptotic-behavior. *Physical Review A, 38*, 3098–3100.
19. Burke, K., Perdew, J., & Wang, Y. (1998). *Electronic density functional theory: recent progress and new directions*. Berlin: Springer.
20. Lee, C., Yang, W., & Parr, R. (1988). Development of the colle-salvetti correlation-energy formula into a functional of the electron-density. *Physical Review B, 37*, 785–789.
21. Vosko, S., Wilk, L., Nusair, M. (1980). Accurate spin-dependent electron liquid correlation energies for local spin-density calculations—a critical analysis. *Canadian Journal of Physics, 58*, 1200–1211.
22. MacKerell, A., Bashford, D., Bellott, M., Dunbrack, R., Evanseck, J., & Field, M., et al. (1998). All-atom empirical potential for molecular modeling and dynamics studies of proteins. *Journal of Physical Chemistry B, 102*, 3586–3616.

Chapter 3
Degrees of Freedom in Polypeptides and Proteins

3.1 Introduction

Proteins are biological polymers consisting of amino acids whose number may vary in the range from several tens up to tens of thousands. Small fragments of proteins are usually called polypeptide chains or polypeptides. This chapter is devoted to a study of the conformational properties of alanine and glycine polypeptide chains.

Recently, it became possible to study small fragments of proteins and polypeptides in the gas phase experimentally with the use of the MALDI mass spectroscopy [1–4] and the ESI mass spectroscopy [5, 6]. From theoretical viewpoint, investigation of small polypeptides is of significant interest because they can be treated by means of ab initio methods which allow accurate comparison of theoretical predictions with experiment. The results of ab initio calculations can be utilized for the development of model approaches applicable for the description of larger and more complex protein structures.

Polypeptides are characterized by primary and secondary structures [7–10]. Different geometrical configurations of polypeptides are often called as the conformations. The number of various conformations (isomeric states) grows rapidly with the growth of a system size. Thus, a search for the most stable conformations becomes an increasingly difficult problem for large molecules.

The sheet and the helix structures are the most abundant motifs in proteins. Study of the transition between these motifs and the evaluation of the characteristic duration of transition in these structures is of significant interest, because it is closely related to one of the most intriguing problems of the protein physics, the protein folding. In order to study this transition it is necessary to investigate the potential energy surface of amino acid chains with respect to their twisting. This chapter of the thesis is devoted to the study of the potential energy surfaces for small alanine and glycine chains. These molecules were chosen because they are often present in native proteins as fragments and also because they allow for an ab initio theoretical treatment due to their relatively small size.

A. V. Yakubovich, *Theory of Phase Transitions in Polypeptides and Proteins,*
Springer Theses, DOI: 10.1007/978-3-642-22592-5_3,
© Springer-Verlag Berlin Heidelberg 2011

Previously, only glycine and alanine dipeptides were studied in detail. Sometimes their analogues were used to reduce the computational costs (for example, (S)-α-(formylamino)propanamide). In Refs. [11–13] alanine and glycine dipeptides were investigated within the Hartree–Fock theory. In these papers, the potential energy surfaces were calculated versus the twisting angles of the molecules. Different stable states of the dipeptides, corresponding to different molecular conformations, were determined. In Refs. [14–19], different conformations and their energies were determined within the framework of the density functional theory. In Ref. [19], the dynamics of the alanine dipeptide analog was discussed and the time of the transitions between the two conformations of the alanine dipeptide was found.

A number of papers were devoted to the study of tripeptides. In Refs. [20–24] dynamics of the alanine and glycine tripeptides was studied by means of classical molecular dynamics and with the use of semi-empirical potentials (such as GROMOS, CHARMM and AMBER). In [25], within the framework of the Hartree-Fock theory, several stable conformations of alanine and glycine tripeptides were found. In Ref. [26], the Raman and IR spectra for alanine and glycine tripeptides were measured in neutral, acidic and alkali environments.

A few works were devoted to the study of polypeptides of greater length. In particular, stable conformations of neutral and charged alanine hexapeptides were obtained with the use of empirical potentials and discussed in ref. [27]. Experimental NMR study of various conformations of alanine heptapeptides at different temperatures was carried out in ref. [27]. In ref. [29], with the use of empirical molecular dynamics based on Monte-Carlo methods, a polypeptide consisting of 21 amino acids was described.

In this chapter, ab initio calculations of the multidimensional potential energy surface for the alanine and glycine polypeptide chains, consisting of three and six amino acids, are discussed . The potential energy surface as function of twisting degrees of freedom of the polypeptide chain has been calculated. The calculations have been performed within ab initio theoretical framework based on the density functional theory (DFT) accounting for all the electrons in the system. Previously, this kind of calculations were performed only for dipeptides (see, e.g., [11, 12, 19]). For larger molecules, only a few conformations were considered (see citations above). In the present work the most energetically favorable conformations of the polypeptides and the energy barriers for the transitions between these conformations are determined.

Using a thermodynamic approach, the times of the characteristic transitions between the most energetically favorable conformations were estimated. The results of the calculations have been compared with other theoretical simulations and with the available experimental data. The influence of the secondary structure on the potential energy landscapes is analyzed as well. In particular, the role of the secondary structure in the formation of stable conformations of the chains of six amino acids being in the sheet and in the helix conformations has been elucidated. The results of the work presented in this chapter are published in [30–33].

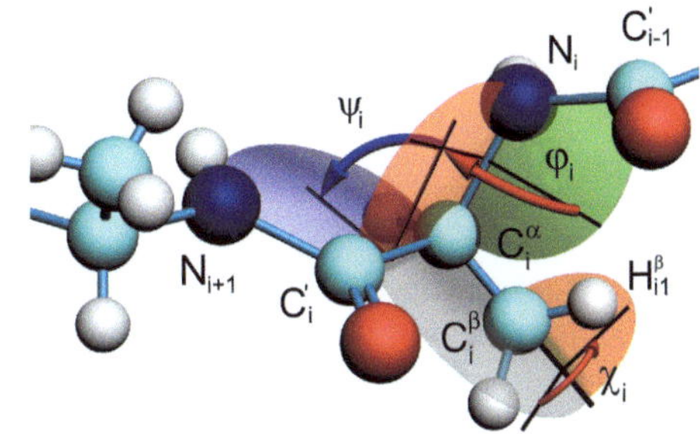

Fig. 3.1 Dihedral angles φ and ψ used for characterization of the secondary structure of a polypeptide chain. The dihedral angle χ_i characterizes the rotation of the side radical along the $C^\alpha_i - C^\beta_i$ bond. The figure is adopted from [34]

3.2 Conformational Properties of Alanine and Glycine Chains

3.2.1 Determination of the Polypeptides Twisting Degrees of Freedom

In this section are presented the potential energy surfaces for the alanine and glycine polypeptide chains calculated versus dihedral angles φ and ψ defined in Fig. 3.1. In particular, the chains consisting of three and six amino acids are considered.

Both angles are defined by the four neighboring atoms in the polypeptide chain. The angle φ_i is defined as the dihedral angle between the planes formed by the atoms $(C'_{i-1} - N_i - C^\alpha_i)$ and $(N_i - C^\alpha_i - C'_i)$. The angle ψ_i is defined as the dihedral angle between the $(N_i - C^\alpha_i - C'_i)$ and $(C^\alpha_i - C'_i - N_{i+1})$ planes. The angle χ_i is defined as the dihedral angle between the planes formed by the atoms $(C'_i - C^\alpha_i - C^\beta_i)$ and by the bonds $C^\alpha_i - C^\beta_i$ and $C^\beta_i - H^\beta_{i1}$. Beside the angles φ_i, ψ_i and χ_i there is an angle ω_i, which is defined as the dihedral angle between $(C^\alpha_i - C'_i - N_{i+1})$ and $(C'_i - N_{i+1} - C^\alpha_{i+1})$ planes. The atoms are numbered from the NH_2- terminal of the polypeptide. The angles φ_i, ψ_i and ω_i take all possible values within the interval $[-180°; 180°]$. For the unambiguous definition, the angles φ_i, ψ_i and ω_i are counted clockwise, if one looks on the molecule from its NH_2- terminal (see Fig. 3.1). This way of angle counting is the most commonly used [10].

The angles φ_i and ψ_i can be defined for any amino acid in the chain, except the first and the last ones. Below the subscripts are omitted and the angles φ and ψ are considered for the middle amino acid of the polypeptide.

3.2.2 Optimized Geometries of Alanine Polypeptides

In order to study twisting of a polypeptide chain one needs first to define its initial structure. Although, the number of its conformations increases with the growth of the molecule size, there are certain types of polypeptide structure, namely the sheet and the helix conformations, which are the most typical. Therefore, the twisting of the

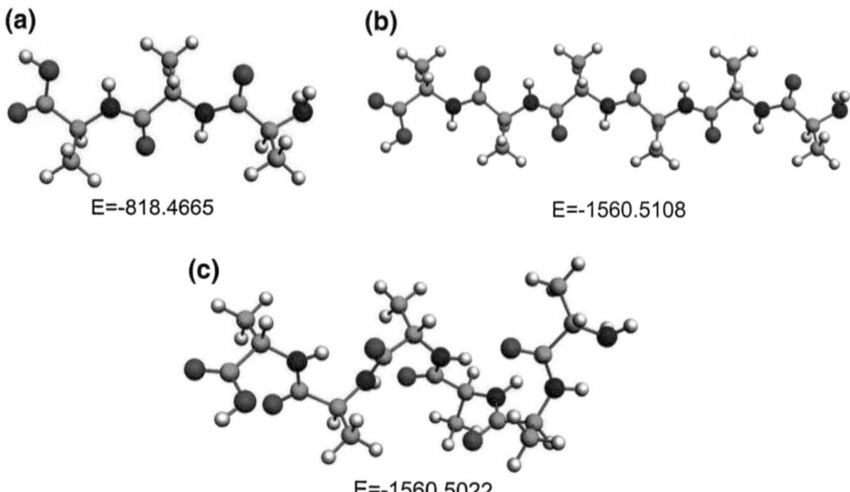

Fig. 3.2 Optimized geometries of alanine polypeptide chains calculated by the B3LYP/ 6-31++G(d,p) method: **a** Alanine tripeptide, **b** Alanine hexapeptide (sheet conformation), **c** Alanine hexapeptide (helix conformation). The figure is adopted from [32]

polypeptide chains of the sheet and the helix conformations is discussed in this work. By varying the angles φ and ψ in the central amino acid one can create the structure of the polypeptide differing significantly from the pure sheet or helix conformations. If the structure of a polypeptide can be transformed to a helix or a sheet one by a trivial variation of φ and ψ, such polypeptides for the sake of simplicity are referred below as belonging to the group of the helix or the sheet structure, respectively.

In Fig. 3.2 are presented the optimized geometries of alanine polypeptide chains that have been used for the exploration of the potential energy surfaces. All geometries were optimized with the use of the B3LYP functional. Figure 3.2a shows the alanine tripeptide structure. In the present work the sheet conformation is chosen, because the tripeptide is too short to form the helix conformation. Figure 3.2b, c show alanine hexapeptide in the sheet and the helix conformations, respectively. The total energies (in atomic units) of the molecules are given below the images.

3.2.3 Polypeptide Energy Dependance On the Dihedral Angle ω

For each amino acid there are only three dihedral angles formed by atoms of the polypeptide chain which describe its twisting. The angle ω (see Fig. 3.1) differs from the angles φ and ψ, because C_i' atom has the sp^2 hybridization state, what leads to formation of a quasi-double bond between C_i' and N_{i+1} atoms. Therefore, the angle ω is often referred as a "stiff" degree of freedom, whose value depends only slightly on both the polypeptide constituent amino acids and the values of other degrees of

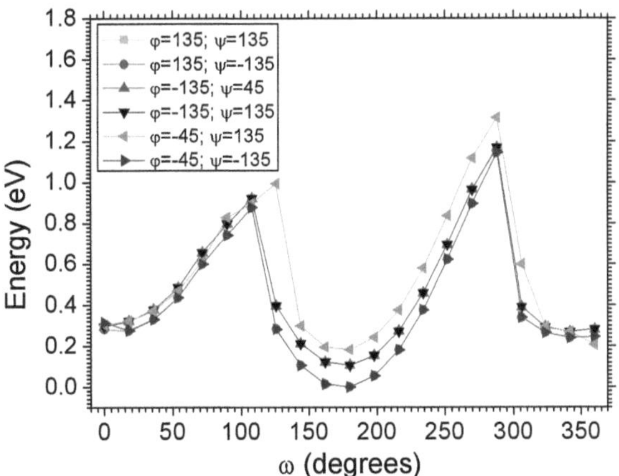

Fig. 3.3 Dependance of alanine tripeptides energy on angle ω calculated by the B3LYP/6-31G(d) method at different values of angles φ and ψ. The figure is adopted from [32]

freedom. To illustrate this fact, in Fig. 3.3 are presented the energy dependencies on ω calculated for alanine dipeptide with different values of angles φ and ψ in the central amino acid.

From this figure it is clear that there are two stable states in the system with $\omega = 0°$ and $\omega = 180°$ which do not depend on the angles φ and ψ. The heights of the barriers between these states are weakly depend on φ and ψ, being equal to $\sim$1 eV $= 23.06$ kcal/mol.

The calculation shows that at temperatures close to the room temperature, the value of the angle ω changes insignificantly. The potential energy surface as a function of the angles φ and ψ appears to be much more complex as it is shown in the next section.

3.2.4 Potential Energy Surface for Alanine Tripeptide

In Fig. 3.4 is presented the potential energy surface for the alanine tripeptide calculated by the B3LYP/6-31G(2d,p) method. The energy scale is given in eV, kcal/mol and Kelvin. Energies on the plot are measured from the lowest energy minimum of the potential energy surface.

From the figure follows that there are several minima on the potential energy surface. They are numbered according to the value of the corresponding energy value. Each minimum corresponds to a certain conformation of the molecule. These conformations differ significantly from each other. In the case of alanine tripeptide there are six conformations, shown in Fig. 3.5. Dashed lines show the strongest hydrogen bonds in the system, which arise when the distance between hydrogen and oxygen atoms becomes less then 2.9 angstroms.

To calculate the potential energy surface the following procedure was adopted. Once the stable structure of the molecule has been determined and optimized, all but two (these are the angles φ and ψ in the central amino acid) degrees of freedom

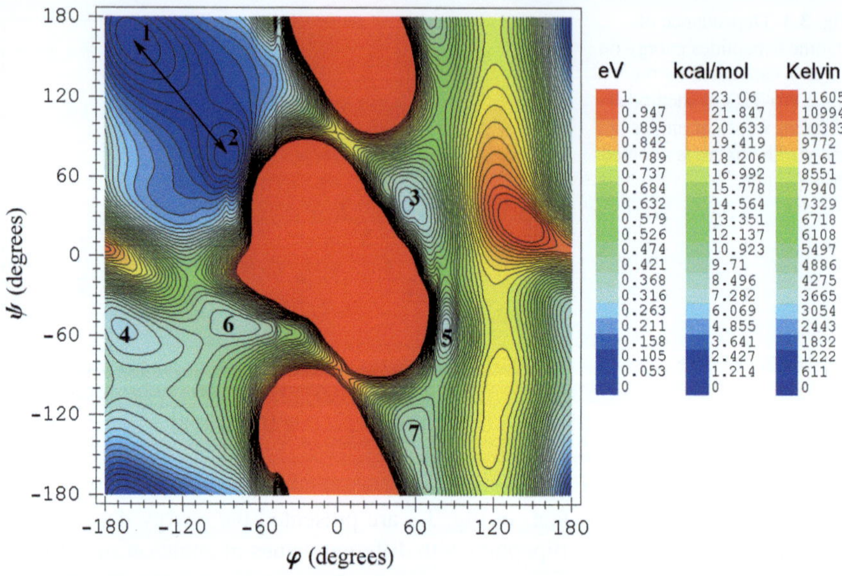

Fig. 3.4 Potential energy surface for the alanine tripeptide calculated by the B3LYP/6-31G(2d,p) method. Energies are given in eV, kcal/mol and Kelvin. Numbers mark energy minima on the potential energy surface. *Arrows* show transition paths between different conformations of the molecule. The figure is adopted from [32]

were frozen. Then the energy of the molecule was calculated by varying φ and ψ. This procedure was used to calculate all potential energy surfaces presented below in this section. It allows one to find efficiently the minima on the energy surface and to determine the main stable conformations of the molecule. The absolute energy values of different conformations of the tripeptide found by this method are not too accurate, because the method does not account for the relaxation of other degrees of freedom in the system. To calculate the potential energy surface with accounting for the relaxation one needs 20–30 times more of the computer time. Therefore, a calculations with accounting for the relaxation have not been performed in this work. Instead, a complete optimization of the molecular conformations, corresponding to all minima on the calculated potential energy surface was performed.

In Fig. 3.5 are compared stable conformations of the alanine tripeptide calculated with and without accounting for the relaxation of all atoms in the system. As it is seen from this figure the angles φ and ψ differ by about 10%in the two cases. This difference arises due to the coupling of φ and ψ with other degrees of freedom. Note the change of the sign of the relative energies of some conformations. This effect is due to the rearrangement of side atoms (radicals) in the polypeptide chain which lowers the energies of different conformations differently.

The potential energy surface has been calculated and interpolated on the grid with the step of $18°$. This step size is an optimal one, because the interpolation error is

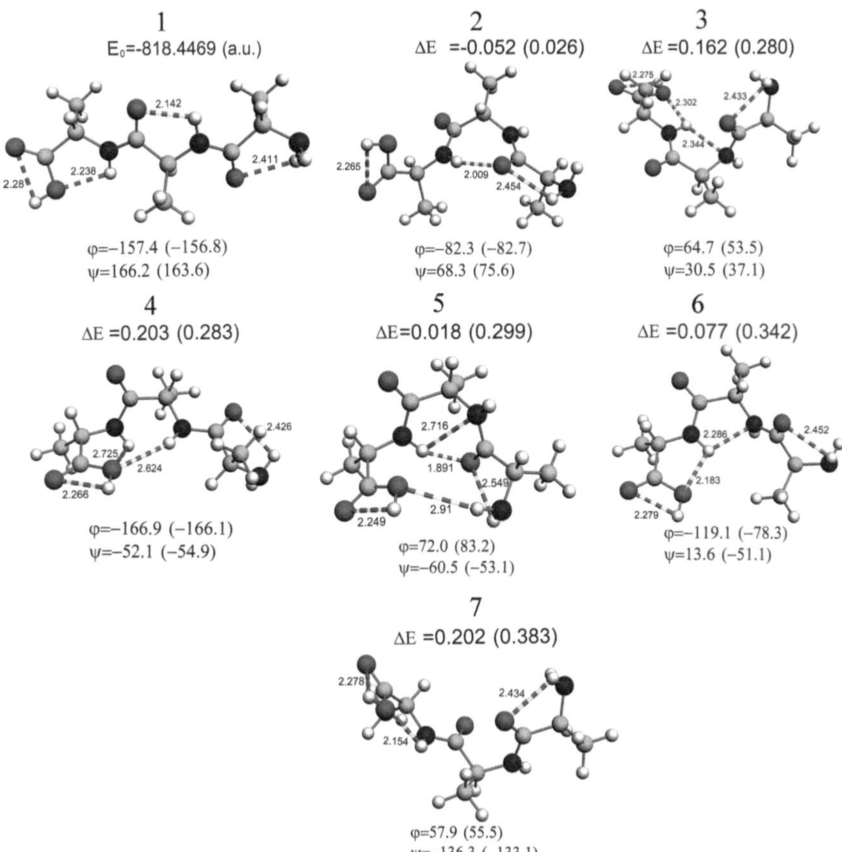

Fig. 3.5 Optimized conformations of the alanine tripeptide. Different geometries correspond to different minima on the potential energy surface (see contour plot in Fig. 3.4. Below each image the angles φ and ψ are presented, which have been obtained with accounting for relaxation of all degrees of freedom in the system. Values in *brackets* give the angles calculated without accounting for relaxation. Above each image the energy of the corresponding conformation is given in eV. The energies are counted from the energy of conformation 1 (the energy of conformation 1 is given in a.u.). Values in *parentheses* correspond to the energies obtained without relaxation of all degrees of freedom in the system. *Dashed lines* show the strongest hydrogen bonds. Their lengths are given in angstroms. The figure is adopted from [32]

about $9°$, i.e. comparable with the angle deviations caused by the relaxation of all degrees of freedom in the system.

Note that for the alanine tripeptide an additional maximum appears at $\varphi = 120° \pm 50°$, $\psi = 30° \pm 30°$, while it is absent on the potential energy surface for the glycine tripeptide (See Fig. 3.12). This maximum is a result of overlapping of the side CH_3- radicals, which are substituted in the case of the glycine polypeptide with the H-atoms.

Table 3.1 Comparison of dihedral angles φ and ψ corresponding to different conformations of alanine tripeptide

Conformation	φ [60]	ψ [60]	φ [61]	ψ [61]	φ	ψ
1	−168.4	170.5	−157.2	159.8	−157.4	166.2
2	−	−	−60.7	−40.7	−82.3	−68.3
3	63.8	32.7	67.0	30.2	64.7	30.5
4	−	−	−	−	−166.9	−52.1
5	74.1	−57.3	76.0	−55.4	72.0	−60.5
6	−128.0	29.7	−130.9	22.3	−119.1	13.6
7	−	−	−	−	57.9	−136.3

In ref. [11] and ref. [12] several stable conformations were found for alanine and glycine dipeptides. The values of angles φ and ψ for the stable conformations of dipeptide and tripeptide are close indicating that the third amino acid in tripeptide makes relatively small influence on the values of dihedral angles of two other amino acids. In earlier papers refs. [11, 12] dipeptides were studied within the framework of the Hartree–Fock theory. In ref. [11], values of φ and ψ were obtained by the HF/6-31+G* method, and in ref. [12] by HF/6-31G**. In Table 3.1 are compared the results of the calculation for tripeptide with the corresponding data obtained for dipeptides. Some discrepancy between the values presented is due to the difference between the dipeptide and tripeptide (i.e. the third alanine in the tripeptide affects the values of angles φ and ψ). However, another source of discrepancy might arise due to accounting for the many-electron correlations in the DFT and neglecting this effect in the Hartree-Fock theory used in refs. [11, 12].

Figure 3.4 shows that some domains of the potential energy surface, where the potential energy of the molecule increases significantly, appear to be unfavorable for the formation of a stable molecular configuration. The growth of energy takes place when some atoms in the polypeptide chain approach each other at small distances. Accounting for the molecule relaxation results in the decrease of the system energy in such cases, but the resulting molecular configurations remain unstable. Such domains on the potential energy surface are called as forbidden ones. In Fig. 3.4 one can identify two forbidden regions in the vicinity of the points (0, 0) and (0, 180). At (0, 0) a pair of hydrogen and oxygen atoms approach to the distances much smaller than the characteristic $H - O$ bond length. This leads to a strong interatomic repulsion caused by the exchange interaction of electrons. At (0, 180) the Coulomb repulsion of pair of oxygen atoms causes the similar effect.

Figure 3.4 shows that there are six minima on the potential energy surface for alanine tripeptide. The transition barrier between the conformations $1 \leftrightarrow 2$ is shown in Fig. 3.6. The barrier has been calculated with and without relaxation of the atoms in the system. The corresponding transition path is marked in Fig. 3.4 by an arrow. This comparison demonstrates that accounting for the relaxation significantly lowers the barrier height and influences the relative value of energy of the minima.

Let us now estimate the time needed for a system for the transition from one conformation to another. It can be done using the Arhenius equation, which reads as:

$$\frac{1}{\tau} = \Omega e^{-\frac{\Delta E}{kT}} \qquad (3.1)$$

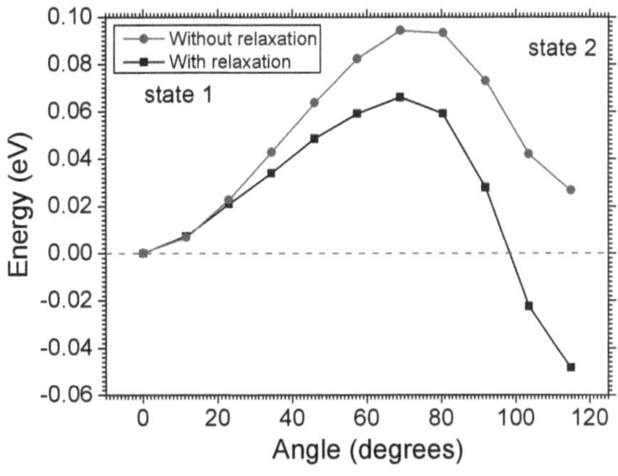

Fig. 3.6 Transition barriers for between conformations $1 \leftrightarrow 2$ of the alanine tripeptide. *Circles* and *squares* correspond to the barriers calculated without and with relaxation of all degrees of freedom in the system. The figure is adopted from [32]

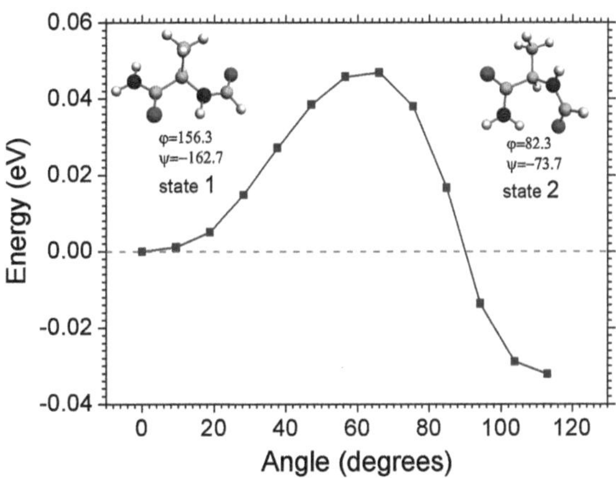

Fig. 3.7 Transition barriers for between conformations $1 \leftrightarrow 2$ of alanine dipeptide analog calculated by the B3LYP/6-31+G(2d,p) method accounting for the relaxation of all degrees of freedom in the system. Structure of the conformations 1 and 2 is shown near each minimum. The figure is adopted from [32]

where τ is the transition time, Ω is the factor, determining how frequently the system approaches the barrier, ΔE is the barrier height, T is the temperature of the system, k is the Boltzmann factor.

Figure 3.7 shows the transition barrier between two main conformations of the alanine dipeptide analog ((S)-α-(formylamino)propanamide). It is seen that $\Delta E_{1 \rightarrow 2} = 0.047$ eV for the transition $1 \rightarrow 2$, while $\Delta E_{2 \rightarrow 1} = 0.079$ eV for the transition $2 \rightarrow 1$. The frequency Ω for this molecule is equal to 42.87 cm^{-1}. Thus, at $T = 300$ K, $\tau^{1 \rightarrow 2}_{2 \times Ala} \sim 5$ ps and $\tau^{2 \rightarrow 1}_{2 \times Ala} \sim 17$ ps. This result is in excellent agreement with the molecular dynamics simulations results obtained in ref. [19] predicting $\tau \sim 7$ ps for the transition $1 \rightarrow 2$ and $\tau \sim 19$ ps for the transition $2 \rightarrow 1$. This comparison demonstrates that the presented method is reliable enough and it can be used for the estimation of transition times between various conformations of the polypeptides.

Using the B3LYP/6-31G(2d,p) method, were calculated the frequencies of normal vibration modes for the alanine tripeptide. The characteristic frequency corresponding to twisting of the polypeptide chain is equal to $32.04\,\text{cm}^{-1}$. From Fig. 3.6 follows that $\Delta E_{1\to2} = 0.066$ eV for the transition $1 \to 2$ and $\Delta E_{2\to1} = 0.114$ eV for the transition $2 \to 1$. Thus, $\tau_{3\times Ala}^{1\to2} \sim 13$ ps and $\tau_{3\times Ala}^{2\to1} \sim 86$ ps. Note, that these transition times can be measured experimentally by means of NMR refs. [10, 35].

3.2.5 Potential Energy Surface for Alanine Hexapeptide with the Sheet and the Helix Secondary Structure

In Fig. 3.8 are presented contour plots of the potential energy surface for the alanine hexapeptide with the sheet (part **a**) and the helix (part **b**) secondary structure, respectively, versus dihedral angles φ and ψ. In both cases the forbidden regions arise because of the repulsion of oxygen and hydrogen atoms analogously to the alanine tripeptide case.

Minima 1–6 on the potential energy surface 3.8a correspond to different conformations of the alanine hexapeptide with the sheet secondary structure. Note that these minima are also present on the potential energy surface of the alanine tripeptide (see Fig. 3.4). Geometries of the conformations 1-6 are shown on the right-hand side of Fig. 3.8a.

Energy barrier as a function of a scan variable (see Fig. 3.8a) for the transition between conformations 1 and 2 is shown in Fig. 3.9. The energy dependence has been calculated with and without relaxation of all the atoms in the system. In the case of alanine hexapeptide with the sheet secondary structure the barrier height for the transition $1 \to 2$ is significantly higher than for the transition $2 \to 1$, being equal to 0.095 and 0.023 eV, respectively. The normal vibration mode frequency, corresponding to the twisting of the polypeptide chain is equal to $6.24\,\text{cm}^{-1}$ and was calculated with the B3LYP/STO-3G method. Using equation (3.1) one derives the transition times at room temperature: $\tau_{6\times Gly}^{1\to2} \sim 211$ ps, $\tau_{6\times Gly}^{2\to1} \sim 13$ ps.

Let us now consider alanine hexapeptide with the helix secondary structure. The potential energy surface for this polypeptide is shown in Fig. 3.8b. The positions of minima on this surface are shifted significantly compared to the cases discussed above. This change takes place because of the influence of the secondary structure of the polypeptide on the potential energy surface. The geometries of the most stable conformations are shown on the right-hand side of Fig. 3.8b.

For the alanine hexapeptide with the helix secondary structure there is a maximum at $\varphi \sim 180°$ and $\psi \sim 40°$ in addition to the central maxima on the potential energy surface. This maximum appears because of the repulsive interaction of the outermost amino acids side radicals.

It is worth noting that for some conformations of alanine hexapeptide the angles φ and ψ change significantly when the relaxation of all degrees of freedom in the system are accounting for (see for example conformations 1, 5 in Fig. 3.8a and conformations 2, 4 in Fig. 3.8b). This means that the potential energy surface of

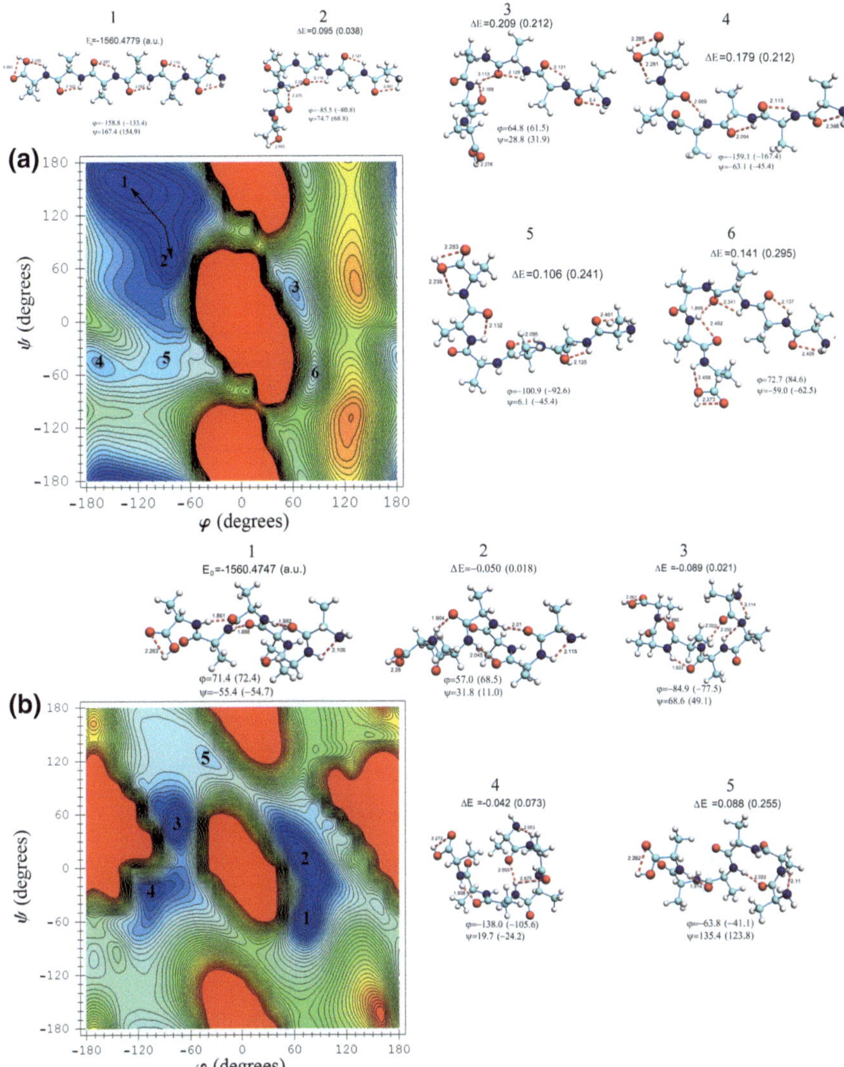

Fig. 3.8 Potential energy surface for the alanine hexapeptide with the sheet secondary structure (part **a**) and with the helix secondary structure (part **b**) calculated by the B3LYP/6-31G(2d,p) method. Energy scale is given in Fig. 3.4. Numbers mark energy minima on the potential energy surface. Images of optimized conformations of the alanine hexapeptide are shown near the corresponding energy landscape. Values of angles φ and ψ, as well as the relative energies of the conformations are given analogously to that in Fig. 3.5. The figure is adopted from [32]

the alanine hexapeptide in the vicinity of the mentioned minima is very sensitive to the relaxation of all degrees of freedom. However, calculation of the potential energy surface with accounting for the relaxation of all degrees of freedom is

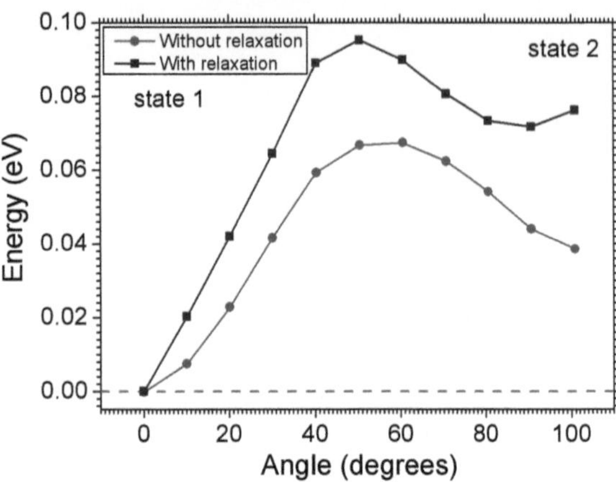

Fig. 3.9 Transitions barriers between conformations $1 \leftrightarrow 2$ of the alanine hexapeptide with the sheet secondary structure. *Circles* and *squares* correspond to the barriers calculated without and with relaxation of all degrees of freedom in the system. The figure is adopted from [32]

unfeasible task. Indeed, one needs about 2,000 h of computer time (Pentium Xeon 2.4 GHz) for the calculation of the potential energy surface for the alanine hexapeptide. To perform an analogues calculation with accounting for the relaxation about 5 years of computer time would be needed. Nevertheless, the potential energy surface calculated without accounting for the relaxation carries a lot of useful information. Thus, one can predetermine stable conformations of polypeptide, which then can be used as starting configurations for further energy minimization.

3.2.6 Comparison of Calculation Results with Experimental Data

Nowadays, the structure of thousands of proteins has been determined experimentally ref. [7]. Knowing the protein structure one can find the angles φ and ψ for each amino acid in the protein.

In Fig. 3.10a is shown a map of the allowed and forbidden conformations for alanine residues in poly-alanine chain taken from ref. [36] (steric Ramachandran diagram). This map was obtained from pure geometrical considerations, in which the structure of the polypeptide was assumed to be fixed and defined by the inter-atomic van der Waals interaction radii. Depending on the distances between the atoms one could distinguish three regions: completely allowed, conventionally allowed and forbidden. The conformation is called completely allowed if all the distances between atoms of different amino acids are larger than some critical value $r_{ij} \geq r_{max}$. Conventionally allowed regions on the potential energy surface correspond to the conformations of the polypeptide, in which the distances between some atoms of different amino acids lie within the interval $r_{min} \leq r_{ij} < r_{max}$. All other conformations are referred to as forbidden. The values of r_{min} and r_{max} are defined by the types of interacting atoms and can be found in the textbooks (see, e.g., [36]). In Fig. 3.10a

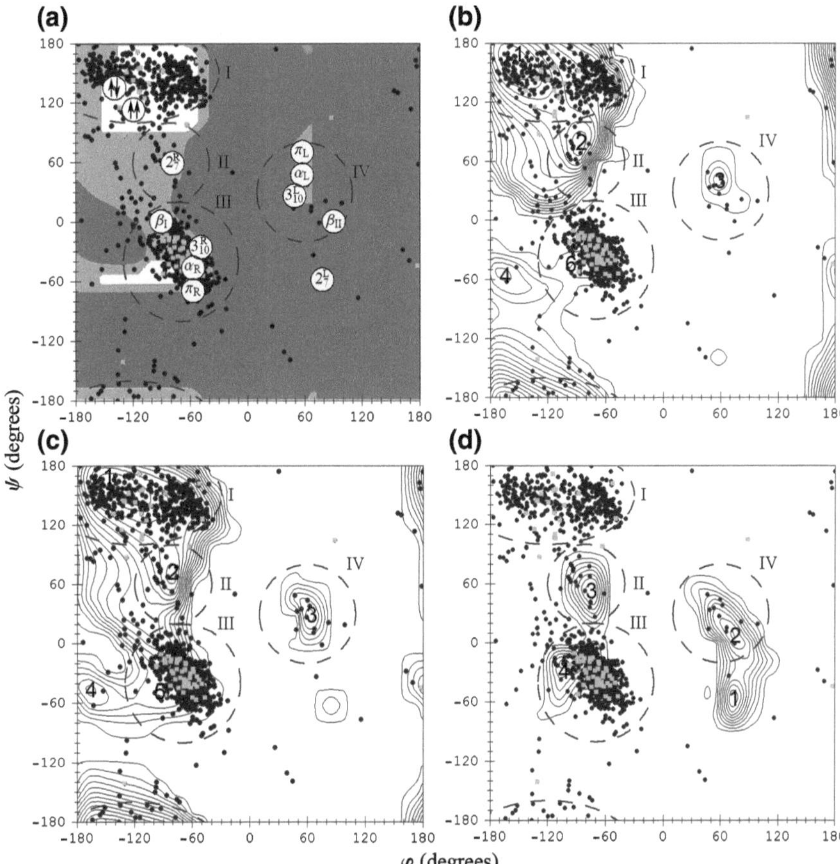

Fig. 3.10 Comparison of angles φ and ψ of alanine residues in protein structures selected from the Brookhaven Protein Data Bank [7, 37] with the steric diagram for poly-alanine [36] (part **a**). Comparison of angles φ and ψ of alanine residues in protein structures selected from the Brookhaven Protein Data Bank [7, 37] with the minima on the calculated potential energy surfaces for: alanine tripeptide (**b**), alanine hexapeptide in sheet conformation (**c**), alanine hexapeptide in helix conformation (**d**). Transparent rhomboids correspond to alanines surrounded with alanines, while filled circles correspond to alanines surrounded by other amino acids. Dashed ellipses mark the regions of higher concentration of the observed angles. The figure is adopted from [32]

the completely allowed regions are marked with white, the conventionally allowed regions with light gray and the forbidden regions with dark gray color. In this figure are marked the points, which correspond to the geometries of alanine, whose periodical iteration leads to the formation of chains with specific secondary structure. In Table 3.2 is presented the compilation of the values of angles φ and ψ, which correspond to the most prominent poly-alanine secondary structures. For the illustrative purposes these points are marked by white circles with the corresponding type of the secondary structure typed in. Thus, 2_7^R, 2_7^L are the right-handed and the left-handed

Table 3.2 Angles φ and ψ corresponding to the most prominent poly-alanine secondary structures

Structure type	φ (Deg.)	ψ (Deg.)
right-handed (left-handed) 2_7 helix	-78 (78)	59 (-59)
right-handed (left-handed) 3_{10} helix	-49 (49)	-26 (26)
right-handed (left-handed) α-helix (4_{13})	-57 (57)	-47 (47)
right-handed (left-handed) π-helix (5_{16})	-57 (57)	-70 (70)
parallel β sheet ($\uparrow\uparrow$)	-119	113
antiparallel β sheet ($\uparrow\downarrow$)	-139	135
β-turn of type I	-90	0
β-turn of type II	90	0

2_7 helix; 3_{10}^R, 3_{10}^L are the right-handed and the left-handed 3_{10} helix; α_R, α_L are the right-handed and the left-handed α-helix (4_{13}); π_R, π_L are the right-handed and the left-handed π-helix (5_{16}); $\uparrow\uparrow$, $\uparrow\downarrow$ are the parallel and antiparallel β sheets. β_I, β_{II} correspond to the β-turns of types I and II, respectively.

Note that not all of the structures listed above are present equally in proteins. In Fig. 3.10a is shown the distribution of the angles φ and ψ of alanine residues in protein structures selected from the Brookhaven Protein Data Bank [7, 37]. It is possible to distinguish four main regions, in which most of experimental points are located. In Fig. 3.10 these regions are schematically shown with *dashed ellipses*. Note, that these ellipses are used for illustrative purposes only, and serve for a better understanding of the experimental data. The regions in which most of the observed angles φ and ψ are located correspond to different secondary structures of the poly-alanine. Thus, region I corresponds to the parallel and antiparallel β-sheets. Region II corresponds to the right-handed 2_7^R helix. Region III corresponds to the right-handed α_R-helix, right-handed π_R-helix, right-handed 3_{10}^R helix and β-turn of type I. Region IV corresponds to the left-handed α_L-helix, right-handed π_L-helix, left-handed 3_{10}^L helix and β-turn of type II. In some cases there are several types of secondary structure within one domain.

Let us now compare the distribution of angles φ and ψ experimentally observed for proteins with the potential energy landscape calculated for alanine polypeptides and establish correspondence of the secondary structure of the calculated conformations with the predictions of the simple Ramachandran model.

Region I corresponds to the minimum 1 on the both potential energy surfaces of the alanine tripeptide (Fig. 3.10b) and the alanine hexpeptide with the secondary structure of sheet (Fig. 3.10c). These conformations correspond exactly to the alanine chains in the β-sheet conformation (see Figs. 3.5 and 3.8a). Note that there is no minimum in that region of the potential energy surface for alanine hexapeptide with the secondary structure of helix (see Fig. 3.10d).

Region II corresponds to the minimum 2 on the both potential energy surfaces 10b and 10c, as well as to the minimum 3 on the potential energy surface 10d. On the steric diagram for poly-alanine this region corresponds to the right-handed 2_7^R helix. The structure of conformations 2 on the surfaces 10b and 10c differs from the structure of this particular helix type. Only the central alanines, for which the angles φ and ψ in Fig. 3.10b, c are defined, have the structure of 2_7^R helix. Thus, one can refer to the conformations 2 as to the mixed states, where the central part of the polypeptide chain has the conformation of helix and the outermost parts have the conformation of sheet. Conformation 3 on the surface 10d is also a mixed state. Here one can distinguish one turn of 3_{10}^R helix and two turns of 2_7^R helix (see Fig. 3.8b).

Region III corresponds to the structure of right-handed α_R-helix, right-handed 3_{10}^R helix, right-handed π_R-helix and β-turn. It corresponds to minima 6, 5 and 4 on the potential energy surfaces 10b, 10c and 10d, respectively. Conformation 6 can not be assigned to any specific type of secondary structure because the chain is too short. Note, that conformation 6 is even not a stable one on the potential energy surface of the alanine tripeptide. The most probable types of secondary structures in that region of the potential energy surface are right-handed α_R-helix and β-turn. However, for the formation of a single turn of α_R-helix (or for the formation of β-turn) at least four amino acids are needed. Conformation 5 on the potential energy surface of the alanine hexapeptide can be characterized as a partially formed β-turn because the alanine, for which the dihedral angles φ and ψ in Fig. 3.10c are defined has the geometry of β-turn, but its neighbor forms a β- sheet (see Fig. 3.8a). Conformation 4 on the potential energy surface 8b changes significantly after accounting for the relaxation of all degrees of freedom in the system, and gets outside the region III. In this conformation one can locate fragments of right-handed 2_7^R and 3_{10}^R helices. The point corresponding to the minimum 4 (after accounting for the relaxation) lies outside regions II and III because angles φ and ψ in Fig. 3.10d are defined for the amino acid between two helix fragments.

Region IV is represented by the structure of left-handed α_L-helix, left-handed 3_{10}^L helix, left-handed π_L-helix and β-turn of type II. The fragments with those types of secondary structures are very rare met in native proteins. To form these structures it is necessary to have at least four amino acids, therefore minima 3 on the potential energy surface for alanine tripeptide can not be compared to any type of the mentioned secondary structures. Region IV corresponds to the conformations 3 and 2 on the surfaces 10c and 10d, respectively. Conformation 3 on the surface 10c corresponds to partially formed β-turn, because the alanine, for which the dihedral angles φ and ψ in Fig. 3.10c are plotted has the configuration of β-turn but the neighboring amino acid in the polypeptide chain forms β- sheet (see Fig. 3.8a). Conformation 2 on the potential energy surface 10d lies outside the region IV, but accounting for the relaxation of all degrees of freedom shifts the minimum on the potential energy surface to the allowed region of left-handed α_L- and 3_{10}^L helix (see Fig. 3.8b). The geometry of conformation 2 is similar to the geometry of left-handed 3_{10}^L helix (see Fig. 3.8b). The main differences in the structure are caused by the insufficient length of the polypeptide chain to form a regular helix structure.

3.3 Conformational Changes in Glycine Tri- and Hexa-Peptide

In this section the potential energy surfaces of glycine polypeptides are discussed. It is not feasible to study the dependence of potential energy on all possible angles φ and ψ for all amino acids, because the amount of computer powers required for

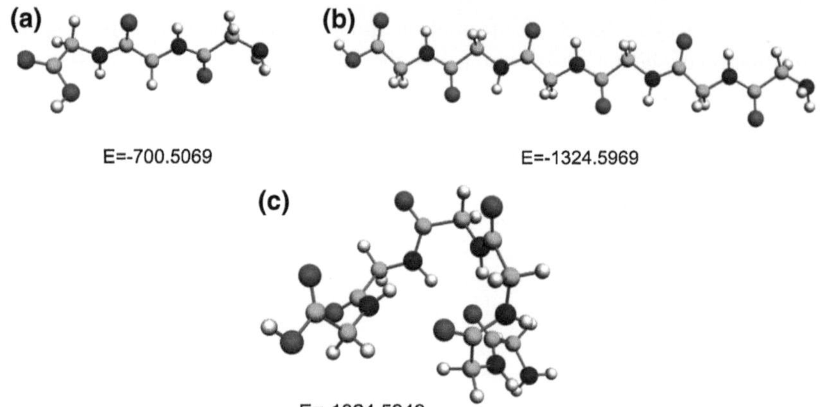

Fig. 3.11 Optimized geometries of glycine polypeptide chains calculated by the B3LYP/6-31++G(d,p) method: **a** Glycine tripeptide; **b** Glycine hexapeptide (sheet conformation), **c** Glycine hexapeptide (helix conformation). The figure is adopted from [30]

such DFT computation would be enormous. Therefore, only the twisting angles in the middle amino acid of the polypeptide are considered, in order to to stress the topological conformity of the potential energy surfaces in the tripeptide and in the hexapeptide. One can expect that for the inner amino acids of a polypeptide the dependence of the potential energy surfaces on the twisting angles should be similar if the amino acids are loosely correlated. For glycines this condition is fulfilled, because glycines do not have side radicals and thus interact weakly with each other along the polypeptide chain.

3.3.1 Optimized Geometries of Glycine Polypeptides

In Fig. 3.11 are presented the optimized geometries of glycine polypeptide chains that have been used for the exploration of the potential energy surfaces. All geometries have been optimized with the use of the B3LYP functional. Figure 3.11a shows the glycine tripeptide structure. In the present work the sheet conformation is chosen, because the tripeptide is too short to form the helix conformation. Figure 3.11b and c show glycine hexapeptide in the sheet and the helix conformations respectively. The total energies (in atomic units) of the molecules are given below the images.

3.3.2 Potential Energy Surface for Glycine Tripeptide

In Fig. 3.12 are presented the potential energy surface for the glycine tripeptide calculated by the B3LYP/6-31G(2d,p) method. The energy scale is given in eV, kcal/mol and Kelvin. Energies on the plot are measured from the lowest energy minimum of the potential energy surface.

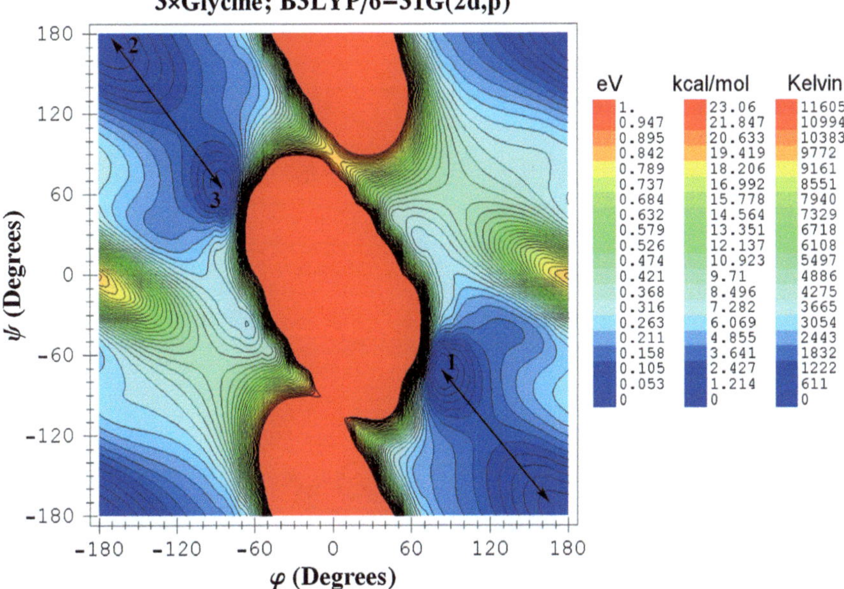

Fig. 3.12 Potential energy surface for the glycine tripeptide calculated by the B3LYP/6-31G(2d,p) method. Energies are given in eV, kcal/mol and Kelvin. Numbers mark energy minima on the potential energy surface. *Arrows* show the transition paths between different conformations of the molecule. The figure is adopted from [30]

From the figure follows that there are several minima on the potential energy surface. They are numbered according to the value of the corresponding energy value. Each minimum corresponds to a certain conformation of the molecule. These conformations differ significantly from each other. In the case of glycine tripeptide there are only three conformations, shown in Fig. 3.13. Dashed lines show the strongest hydrogen bonds in the system, which arise when the distance between hydrogen and oxygen atoms becomes less then 2.9 angstroms.

To calculate the potential energy surface was adopted the same procedure as for alanine polypeptides (Sect. 3.2.4).

In Table 3.3 the results of the calculation for tripeptide are compared with the corresponding data obtained for dipeptides. Some discrepancy between the values presented is due to the difference between the dipeptide and tripeptide (i.e. the third glycine in tripeptide affects the values of angles φ and ψ). However, another source of discrepancy might arise due to accounting for the many-electron correlations in the DFT and neglecting this effect in the Hartree–Fock theory used in refs. [11, 12].

Figure 3.12 shows that some domains of the potential energy surface, where the potential energy of the molecule increases significantly, appear to be unfavorable for the formation of a stable molecular configuration.

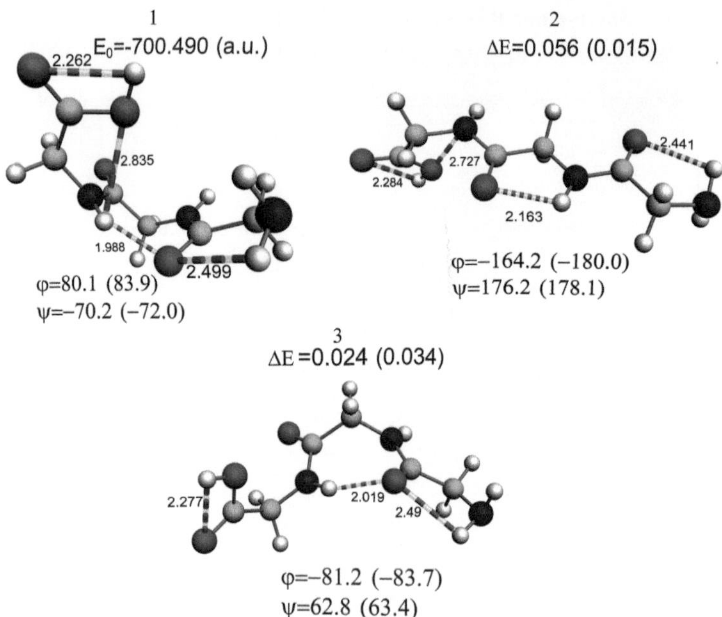

Fig. 3.13 Optimized conformations of the glycine tripeptide. Different geometries correspond to different minima on the potential energy surface (see contour plot in Fig. 3.12). Below each image are presented angles φ and ψ, which have been obtained with accounting for relaxation of all degrees of freedom in the system. Values in *brackets* give the angles calculated without accounting for relaxation. Above each image, the energy of the corresponding conformation is given in eV. The energies are counted from the energy of conformation 1 (the energy of conformation 1 is given in a.u.). Values in *brackets* give the energies obtained without accounting for the relaxation of all degrees of freedom in the system. *Dashed lines* show the strongest hydrogen bonds. Their lengths are given in angstroms. The figure is adopted from [30]

Table 3.3 Comparison of dihedral angles φ and ψ corresponding to different conformations of glycine tripeptide (column 3) with angles φ and ψ for glycine dipeptide from ref. [11, 12] (column 1 and 2)

conformation	φ, ref. [11]	ψ, ref. [11]	φ, ref. [12]	ψ, ref. [12]	φ	ψ
1	–	–	76.0	−55.4	80.1	−70.2
2	−180.0	180.0	−157.2	159.8	−164.2	176.2
3	−85.2	67.4	−85.8	79.0	−81.2	62.8

At (0, 0) a pair of hydrogen and oxygen atoms approach to the distances much smaller than the characteristic $H - O$ bond length. This leads to a strong interatomic repulsion caused by the exchange interaction of electrons. At (0, 180) the Coulomb repulsion of pair of oxygen atoms causes the similar effect.

Figure 3.12 shows that there are three minima on the potential energy surface for glycine tripeptide. The transition barriers between the conformations $2 \leftrightarrow 1$ and $2 \leftrightarrow 3$ are shown in Fig. 3.14. They have been calculated with and without relaxation of the

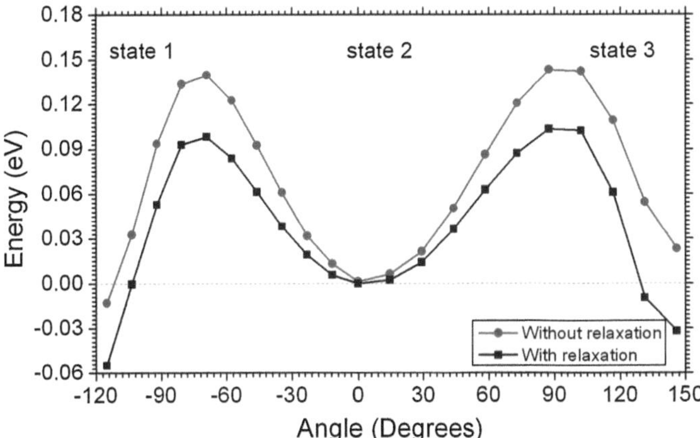

Fig. 3.14 Transition barriers between conformations $1 \leftrightarrow 2 \leftrightarrow 3$ of the glycine tripeptide. *Circles* and *squares* correspond to the barriers calculated without and with relaxation of all degrees of freedom in the system. The figure is adopted from [30]

atoms in the system. The corresponding transition paths are marked in Fig. 3.12 by arrows. This comparison demonstrates that accounting for the relaxation significantly lowers the barrier height and influences the relative value of energy of the minima.

Let us now estimate the time needed for a system for the transition from one conformation to another applying Arhenius equation defined in (3.1). The frequencies of normal vibration modes for the glycine tripeptide were calculated using the B3LYP/6-31G(2d,p) method. The characteristic frequency corresponding to twisting of the polypeptide chain is equal to 33.49 cm^{-1}. From Fig. 3.14 follows that $\Delta E_{2\to3} = 0.103$ eV for the transition $2 \to 3$, and $\Delta E_{3\to2} = 0.132$ eV for the transition $3 \to 2$. Thus, $\tau_{3\times Gly}^{2\to3} \sim 54$ ps and $\tau_{3\times Gly}^{3\to2} \sim 164$ ps. The transition times can be also measured experimentally using NMR technique [10, 35].

3.3.3 Potential Energy Surface for Glycine Hexapeptide with the Sheet and the Helix Secondary Structure

In Fig. 3.15 are presented contour plots of the potential energy surfaces for the glycine hexapeptide with the sheet (part a) and the helix (part b) secondary structure, respectively, versus dihedral angles φ and ψ. In both cases the forbidden regions arise because of the repulsion of oxygen and hydrogen atoms analogously to the glycine tripeptide case.

Minima 1–5 on the potential energy surface 15a correspond to different conformations of the glycine hexapeptide with the sheet secondary structure. Note that minima 1–3 are also present on the potential energy surface of the glycine tripeptide. Geometries of the conformations 1–5 are shown on the right-hand side of Fig. 3.15a.

For the glycine hexapeptide with the sheet secondary structure additional minima 4–5 arise. The appearance of these minima is the result of the interaction of the outermost amino acids, which are absent in the case of tripeptide.

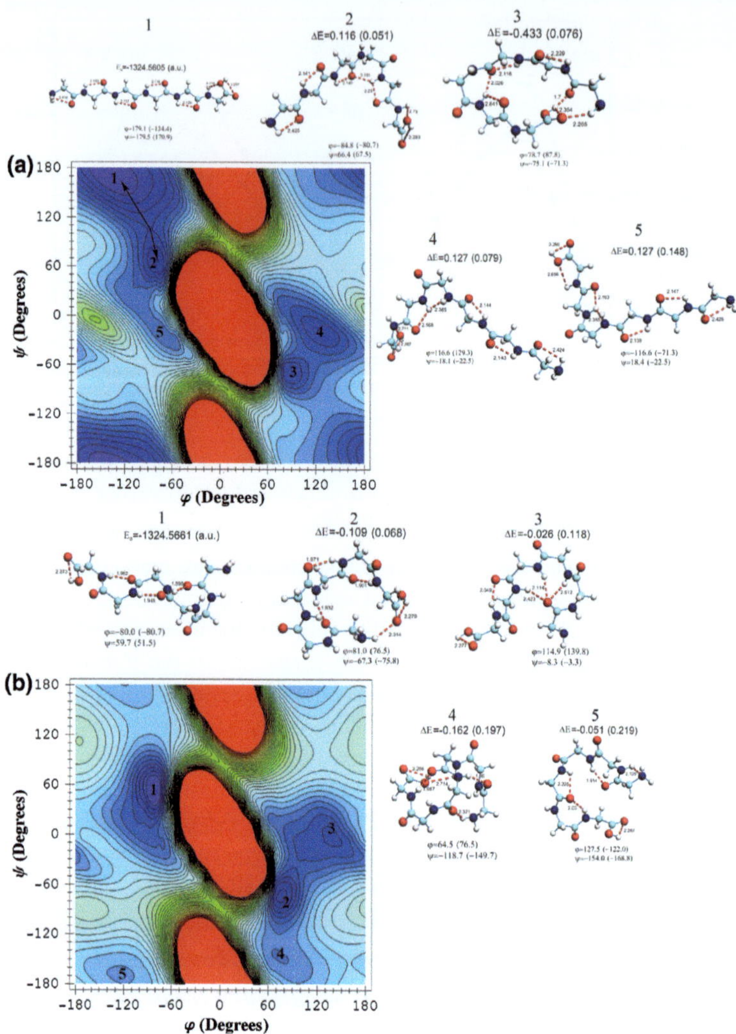

Fig. 3.15 Potential energy surface for the glycine hexapeptide with the sheet secondary structure (part **a**) and with the helix secondary structure (part **b**) calculated by the B3LYP/6-31G(2d,p) method. Energy scale is given in Fig. 3.12. Numbers mark energy minima on the potential energy surface. Images of optimized conformations of the glycine hexapeptide are shown near the corresponding energy landscape. Values of angles φ and ψ, as well as the relative energies of the conformations are given analogously to that in Fig. 3.13. The figure is adopted from [30]

Energy barrier as a function of a scan variable (see Fig. 3.15a) for the transition between conformations 1 and 2 is shown in Fig. 3.16. The energy dependence has been calculated with and without relaxation of all the atoms in the system. In the case of glycine hexapeptide with the sheet secondary structure the barrier height (0.128 eV) for the transition $1 \rightarrow 2$ appears to be close to the corresponding barrier height of the glycine tripeptide (0.103 eV), while the barrier height for the transition $2 \rightarrow 1$ is

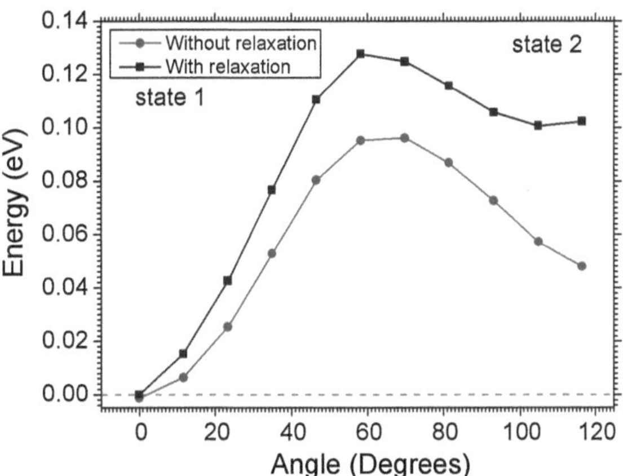

Fig. 3.16 Transition barriers between conformations $1 \leftrightarrow 2$ of the glycine hexapeptide with the sheet secondary structure. *Circles* and *squares* correspond to the barriers calculated without and with relaxation of all degrees of freedom in the system. The figure is adopted from [30]

significantly lower (0.028 eV). The normal vibration mode frequency, corresponding to the twisting of the polypeptide chain is equal to 15.45 cm^{-1} and was calculated with the B3LYP/6-31G(2d,p) method. Using equation (3.1) one derives the transition times at room temperature: $\tau_{6 \times Gly}^{1 \to 2} \sim 305$ ps, $\tau_{6 \times Gly}^{2 \to 1} \sim 6$ ps.

It is well known that the DFT method in its most simple formulation does not reproduce the attractive polarization (van der Waals) interaction well enough, for a discussion see, e.g., Ref. [38] and references therein. However, the polarization interaction can partially be included in the DFT framework via the correlation functional. The degree of accounting for the polarization effects can be controlled by the choice of the number of the polarization functions included in the basis set.

The combination of the B3LYP functional with the 6-31G(2d,p) basis set, which is used for the computations accounts for the polarization interaction on the level of 50 % at least. The most straightforward way of accounting for this energy correction more precisely would be the calculation performed within the framework of the perturbation theory. However, it is not feasible to perform such calculations for molecular systems like hexapeptides, because they would require enormous computer powers.

Another way of accounting for the van der Waals (vdW) interaction is based on the addition of phenomenological Lennard–Jones-type of terms to the total energy of the system [38]. Each of the potentials includes at least two parameters: the equilibrium separation distance of a pair of atoms, and the potential energy well depth. Unfortunately, there are no fixed values for these parameters which would be universally applicable in a wide scope of situations. Even for the same systems different authors choose different parameters (see e.g., [38–40]). Some arbitrariness in the choice of the vdW correction seems to be a general problem of the hybrid approaches.

A systematic study of the potential energy surfaces for polypeptide chains with vdW energy correction taken into account is an interesting problem, which however lies beyond the scope of the thesis. Thus, in the current chapter the vdW interaction energies are calculated only for a few key-points on the potential energy surface, in order to establish the level of accuracy of the calculations. The vdW energy correction for the transition between the conformations 1 and 2 of the glycine hexapeptide is estimated using the set of constants suggested in [39]. The maximum energy value was obtained for the conformation 2, and is equal to 0.0093 eV, which gives the relative error of about 9 %. For the state corresponding to the barrier maximum the vdW energy correction is equal to 0.0058 eV (relative error 4.5 %). Note, that these energies depend strongly on the set of the chosen constants. For example, the vdW energy correction calculated with a set of constants suggested in [40] for conformation 2 is equal to 0.0076 eV, resulting in the relative error of 7.5 %. These estimates demonstrate that the vdW polarization energy plays an important role giving the relative error to the calculated energies of the order of about 5–10 %. Note, that this fact does not necessarily imply changes of the topology of the calculated potential energy surfaces on the same scale, because the relative variations of energies on the potential energy surface are much smaller.

Let us now consider glycine hexapeptide with the helix secondary structure. The potential energy surface for this polypeptide is shown in Fig. 3.15b. The positions of minima on this surface are shifted significantly compared to the cases discussed above. This change takes place because of the influence of the secondary structure of the polypeptide on the potential energy surface. The geometries of the most stable conformations are shown on the right hand-side of Fig. 3.15b.

It is worth noting that for some conformations of glycine hexapeptide the angles φ and ψ change significantly when the relaxation of all degrees of freedom in the system is accounted for (see for example conformations 1, 4, 5 in Fig. 3.15a and conformations 3, 4 in Fig. 3.15b). This means that the potential energy surface of the glycine hexapeptide in the vicinity of mentioned minima is very sensitive to the relaxation of all degrees of freedom. However, calculation of the potential energy surface with accounting for the relaxation of all degrees of freedom is unfeasible task. Indeed, one needs about 1,000 hours of computer time (Pentium Xeon 2.4 GHz) for the calculation of the potential energy surface for the glycine hexapeptide. To perform an analogues calculation with accounting for the relaxation about 3 years of computer time would be needed. Nevertheless, the potential energy surface calculated without accounting for the relaxation carries a lot of useful information. Thus, one can predetermine stable conformations of polypeptide, which then can be used as starting configurations for further energy minimization.

3.3.4 Comparison of Calculation Results with Experimental Data

Nowadays, the structure of many proteins has been determined experimentally [7]. Knowing the protein structure one can find the angles φ and ψ for each amino acid in the protein.

In Fig. 3.17a, a map of the allowed and forbidden conformations for glycine residues in poly-glycine chain is shown. The map is taken from [36] (steric Ramachan-

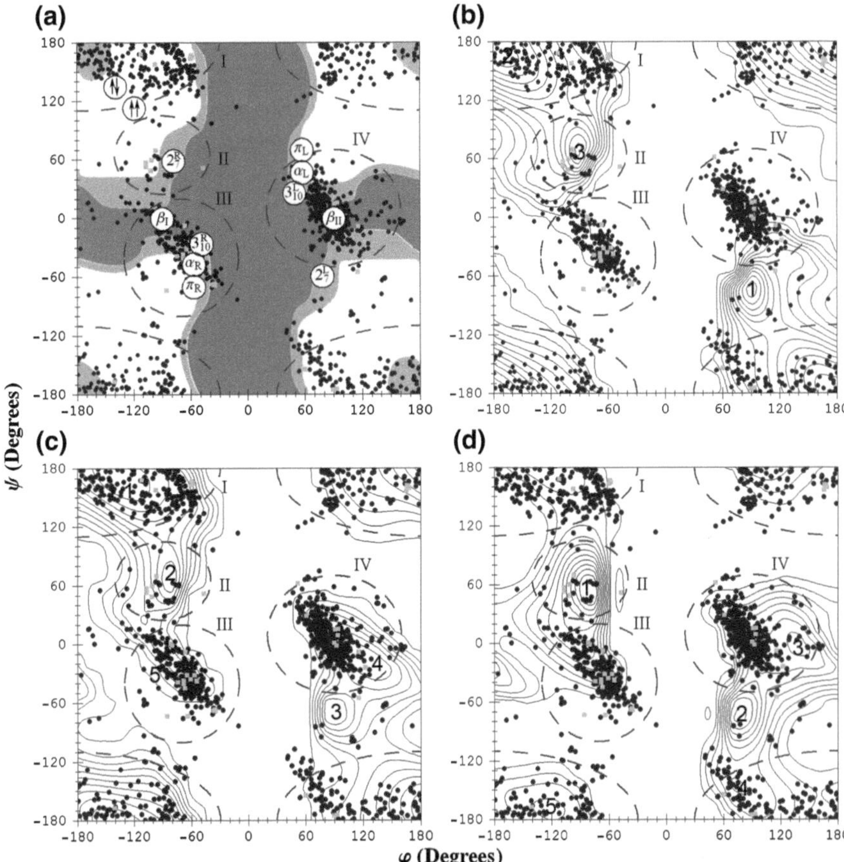

Fig. 3.17 Comparison of angles φ and ψ of glycine residues in protein structures selected from the Brookhaven Protein Data Bank [7, 37] with the steric diagram for poly-glycine [36] (part **a**). Comparison of angles φ and ψ of glycine residues in protein structures selected from the Brookhaven Protein Data Bank [7, 37] with the minima on the calculated potential energy surfaces for: glycine tripeptide (**b**), glycine hexapeptide in sheet conformation (**c**), glycine hexapeptide in helix conformation (**d**). Transparent rhomboids correspond to glycines surrounded with glycines, while *filled circles* correspond to glycines surrounded by other amino acids. *Dashed ellipses* mark the regions of higher concentration of the observed angles. The figure is adopted from [30]

dran diagram). This map was obtained from pure geometrical considerations, in which the structure of the polypeptide was assumed to be fixed and defined by the interatomic van der Waals interaction radii. Depending on the distances between the atoms one could distinguish three regions: completely allowed, conventionally allowed and forbidden. The conformation is called completely allowed if all the distances between atoms of different amino acids are larger than some critical value $r_{ij} \geq r_{max}$. Conventionally allowed regions on the potential energy surface correspond to the conformations of the polypeptide, in which the distances between some

Table 3.4 Angles φ and ψ corresponding to the most prominent poly-glycine secondary structures

Structure type	φ (Deg.)	ψ (Deg.)
right-handed (left-handed) 2_7 helix	-78 (78)	59 (-59)
right-handed (left-handed) 3_{10} helix	-49 (49)	-26 (26)
right-handed (left-handed) α-helix (4_{13})	-57 (57)	-47 (47)
right-handed (left-handed) π-helix (5_{16})	-57 (57)	-70 (70)
parallel β sheet ($\uparrow\uparrow$)	-119	113
antiparallel β sheet ($\uparrow\downarrow$)	-139	135
β-turn of type I	-90	0
β-turn of type II	90	0

atoms of different amino acids lie within the interval $r_{min} \leq r_{ij} < r_{max}$. All other conformations are referred to as forbidden. The values of r_{min} and r_{max} are defined by the types of interacting atoms and can be found in the textbooks (see, e.g., [36]). In Fig. 3.17a the completely allowed regions are marked with white, the conventionally allowed regions with light-gray and the forbidden regions with dark gray color. In this figure the points, which correspond to the geometries of glycine, whose periodical iteration leads to the formation of chains with specific secondary structure are marked. The values of angles φ and ψ, which correspond to the most prominent poly-glycine secondary structures are compiled in Table 3.4 . These points are marked for illustrative purposes by white circles with the corresponding type of the secondary structure typed in. Thus, 2_7^R, 2_7^L are the right-handed and the left-handed 2_7 helix; 3_{10}^R, 3_{10}^L are the right-handed and the left-handed 3_{10} helix; α_R, α_L are the right-handed and the left-handed α-helix (4_{13}); π_R, π_L are the right-handed and the left-handed π-helix (5_{16}); $\uparrow\uparrow$, $\uparrow\downarrow$ are the parallel and antiparallel β sheets. β_I, β_{II} correspond to the β-turns of types I and II respectively.

Note that not all of the structures listed above are present equally in proteins. The distribution of the angles φ and ψ of glycine residues in protein structures selected from the Brookhaven Protein Data Bank [7, 37] is shown in Fig. 3.17. It is possible to distinguish four main regions, in which most of the experimental points are located. In Fig. 3.17 these regions are schematically shown with dashed ellipses. Note that these ellipses are used for illustrative purposes only, and serve for a better understanding of the experimental data. The regions in which most of the observed angles φ and ψ are located correspond to different secondary structure of the poly-glycine. Thus, region I corresponds to the parallel and antiparallel β-sheets. Region II corresponds to the right-handed 2_7^R helix. Region III corresponds to the right-handed α_R-helix, right-handed 3_{10}^R helix, right-handed π_R-helix and β-turn of type I. Region IV corresponds to the left-handed α_L-helix, left-handed 3_{10}^L helix and β-turn of type II. In some cases there are several types of secondary structure within one domain.

Note that some experimental points lie in the forbidden region of the steric Ramachandran diagram (see region IV in Fig. 3.17a). The quantum calculation shows that in fact this region is allowed and has several minima on the potential energy surface (see Figs. 3.17c and d). This comparison shows that for an accurate description of polypeptides it is important to take quantum properties of these systems into account.

Let us now compare the distribution of angles φ and ψ experimentally observed for proteins with the potential energy landscape calculated for glycine polypeptides, and establish correspondence of the secondary structure of the calculated conformations with the predictions of the simple Ramachandran model.

Region I corresponds to the minima 2, 1 and 5 on the potential energy surfaces of the glycine tripeptide (Fig. 3.17b), the glycine hexpeptide with the secondary structure of sheet (Fig. 3.17c), and the glycine hexapeptide with the secondary structure of helix (Fig. 3.17d) respectively. Conformations 2 and 1 in Fig. 3.17b and c correspond exactly to the glycine chains in the β-sheet conformation (see Figs. 3.13 and 3.15a). Conformation 5 in Fig. 3.17d is a mixed state. Here the central amino acid has the conformation of sheet, while the outermost amino acids have the conformation of helix.

Region II corresponds to the minima 3, 2 and 1 on the potential energy surfaces 3.17b, c, and d, respectively. On the steric diagram for poly-glycine (see Fig. 3.17a) this region corresponds to the right-handed 2_7^R helix. The structure of conformations 3 and 2 on the surfaces 3.17b and c differs from the structure of this particular helix type. Only the central glycines, for which the angles φ and ψ in Fig. 3.17b and c are defined, have the structure of 2_7^R helix. Thus, one can refer to conformations 3 and 2 as to the mixed states, where the central part of the polypeptide chain has the conformation of helix and the outermost parts have the conformation of sheet. Conformation 1 on the surface 3.17d is also a mixed state. Here one can distinguish one turn of 3_{10}^R helix and two turns of 2_7^R helix (see Fig. 3.15b).

Region III corresponds to the structure of right-handed α_R-helix, right-handed 3_{10}^R helix, right-handed π_R-helix and β-turn I. It corresponds to minimum 5 on the potential energy surface of the glycine hexapeptide with the secondary structure of sheet 3.17c. Conformation 5 can be characterized as partially formed β-turn, because the glycine, for which the dihedral angles φ and ψ in Fig. (3.17c) are defined, has the geometry of β-turn and its neighbor forms a β-sheet (see Fig. 3.15a). There are no minima in region III on the potential energy surfaces presented in Fig. 3.17b and d. This happens because in this case most probable is the structure of right-handed α_R-helix. To form one turn of the helix of this type it is necessary to link at least four amino acids, so the glycine tripeptide is too short for that. As well, six amino acids chain is too short to form a stable fragment of an α_R-helix, because it does not have enough hydrogen bonds to stabilize the structure. For the hexapeptide more probable are the elements of 3_{10}^R and 2_7^R helixes, because in these cases 2 and 3 helical turns respectively can be formed.

Region IV is represented by the structure of left-handed α_L-helix, left-handed 3_{10}^L helix and β-turn of type II. To form these structures it is necessary to have at least four amino acids, therefore there is no minima in this region on the potential energy surface of glycine tripeptide (Fig. 3.17b). Region IV corresponds to the conformations 4 and 3 on the surfaces 3.17c and d respectively. Conformation 4 on the surface 3.17c corresponds to partially formed β-turn, because the glycine, for which the dihedral angles φ and ψ in Fig. 3.17c are plotted has the configuration of β-turn, but the neighboring aminoacid in the chain forms a β-sheet (see Fig. 3.15a). Conformation 3 on the surface 3.17d can be characterized as deformed turn of left-handed α_L-helix, which turns out to be energetically the most favorable in this region of the potential energy surface (see Fig. 3.15b).

Finally, let us mention a few peculiarities of the calculated potential energy landscapes. At each of the potential energy surfaces discussed in this work one can see a minimum at $\varphi \sim 80°$ and $\psi \sim -70°$. On the steric diagram for poly-glycine this region corresponds to the left-handed $2\frac{L}{7}$ helix. Conformations 1, 3 and 2 on the potential energy surfaces 3.17b, c, and d, respectively, partially represent this structure (see Figs. 3.13, 3.15a and b). The structure of conformation 4 on the potential energy surface 3.17d is similar to left-handed α_L-helix, but differs from it due to the short length of the polypeptide chain, resulting in significant variation of angles φ and ψ in all the residues along the chain.

References

1. Karas, M., & Hillenkamp, F. (1988). Laser desorption ionization of proteins with molecular masses exceeding 10000 daltons. *Analytical Chemistry, 60*, 2299–2301
2. Hillenkamp, F., & Karas, M. (2000). Matrix-assisted laser desorption/ionisation, an experience. *International Journal of Mass Spectrometry, 200*, 71–77.
3. Karas, M., Bahr, U., Fournier, I., Gluckmann, M., & Pfenninger, A. (2003). The initial-ion velocity as a marker for different desorption-ionization mechanisms in maldi. *International Journal of Mass Spectrometry, 226*, 239–248.
4. Wind M., & Lehmann, W. (2004). Element and molecular mass spectrometry - an emerging analytical dream team in the life sciences. *Journal of Analytical Atomic Spectrometry, 19*, 20–25.
5. Fenn, J., Mann, M., Meng, C., Wong, S., & Whitehouse, C. (1989). Electrospray ionization for mass-spectrometry of large biomolecules. *Science, 246*, 64–71.
6. Bröndsted-Nielsen, S., Andersen, J., Hvelplund, P., Liu, B., & Tomita, S. (2004). Biomolecular ions in accelerators and storage rings. *Journal of Physics B: Atomic Molecular and Optical Physics, 37*, 25–56.
7. Protein data bank. (2010). http://www.rcsb.org/
8. Finkelstein A., & Ptitsyn O. (2002). *Protein Physics: A Course of Lectures.* Oxford: Elsevier Books.
9. Mülberg, A. (2004). *Protein Folding.* Russia: St. Petersburg University Press.
10. Rubin, A. (2004). *Biophysics: Theoretical Biophysics.* Moscow: Moscow University Press "Nauka".
11. Head-Gordon, T., Head-Gordon, M., Frisch, M., Brooks, C., & Poplet, J. (1991). Theoretical-study of blocked glycine and alanine peptide analogs. *Journal of American Chemical Society, 113*, 5989–5997
12. Gould, I., Cornell, W., & Hillier, I. (1994). A quantum-mechanical investigation of the conformational energetics of the alanine and glycine dipeptides in the gas-phase and in aqueous-solution. *Journal of American Chemical Society, 116*, 9250–9256.
13. Wang, Z., & Duan, Y. (2004). Solvation effects on alanine dipeptide: A mp2/cc-pvtz//mp2/6-31g** study of (phi, psi) energy maps and conformers in the gas phase, ether, and water. *Journal of Computational Chemistry, 25*, 1699–1716.
14. Perczel A., Farkas O., Jákli I., Topol I., & Csizmadia I. (2003). Peptide models. 13. extrapolation of low-level hartree-fock data of peptide conformation to large basis set scf, mp2, dft, and ccsd(t) results. the ramachandran surface of alanine dipeptide computed at various levels of theory. *Journal of Computational Chemistry, 24*, 1026–1042.

15. Húdaky I., Húdaky P., & Perczel A. (2004). Solvation model induced structural changes in peptides. a quantum chemical study on ramachandran surfaces and conformers of alanine diamide using the polarizable continuum model. *Journal of Computational Chemistry, 25*, 1522–1531
16. Improta, R., & Barone, V. (2004). Assessing the reliability of density functional methods in the conformational study of polypeptides: The treatment of intraresidue nonbonding interactions. *Journal of Computational Chemistry, 25*, 1333–1341
17. Vargas, R., Garza, J., Hay, B., & Dixon, D. (2002). Conformational study of the alanine dipeptide at the mp2 and dft levels. *Journal of Physical Chemistry A, 106*, 3213–3218.
18. Kaschner, R., & Hohl, D. (1998). Density functional theory and biomolecules: a study of glycine, alanine, and their oligopeptides. *Journal of Physical Chemistry A, 102*, 5111–5116.
19. Wei, D., Guo, H., & Salahub, D. (2001). Conformational dynamics of an alanine dipeptide analog An ab initio molecular dynamics study.*Physical Review E, 64*, 011907–(1–14).
20. Woutersen, S., Mu, Y., Stock, G., & Hamm, P. (2001). Hydrogen-bond lifetime measured by time-resolved 2d-IR spectroscopy: N-methylacetamide in methanol. *Chemical Physics, 226*, 131–147.
21. Woutersen, S., Pfister, R., Mu, Y., Kosov, D., & Stock, G. (2002). Peptide conformational heterogeneity revealed from nonlinear vibrational spectroscopy and molecular-dynamics simulations. *Journal of Chemical Physics, 117*, 6833–6840.
22. Mu, Y., & Stock, G. (2002). Conformational dynamics of trialanine in water: a molecular dynamics study. *Journal of Physical Chemistry B, 106*, 5294–5301.
23. Mu, Y., Kosov, D., & Stock, G. (2003). Conformational dynamics of trialanine in water. 2. comparison of amber, charmm, gromos, and opls force fields to nmr and infrared experiments. *Journal of Physical Chemistry B, 107*, 5064–5073.
24. Nguyen, P., & Stock, G. (2003). Nonequilibrium molecular-dynamics study of the vibrational energy relaxation of peptides in water. *Journal of Chemical Physics, 119*, 11350–11358.
25. Torii, H., & Tasumi, M. (1998). Ab initio molecular orbital study of the amide 1 vibrational interactions between the peptide groups in di- and tripeptides and considerations on the conformation of the extended helix. *Journal of Raman Spectroscopy, 29*, 81–86.
26. Schweitzer-Stenner, R., Eker, F., Huang, Q., & Griebenow, K. (2001). Dihedral angles of trialanine in d2o determined by combining ftir and polarized visible raman spectroscopy. *Journal of the American Chemical Society, 123*, 9628–9633.
27. Levy, Y., & Becker, O. (2001). Energy landscapes of conformationally constrained peptides. *Journal of Chemical Physics, 114*, 993–1009.
28. Shi, Z., Olson, C., Rose, G., Baldwin, R., & Kallenbach, N. (2002). Polyproline 2 structure in a sequence of seven alanine residues. *Proceedings of National Academy of Sciences U S A, 99*, 9190–9195.
29. Garcia, A. (2004). Characterization of non-alpha helical conformations in Ala peptides. *Polymer, 45*, 669–676.
30. Yakubovich, A., Solov'yov, I., Solov'yov, A., & Greiner, W. (2006). Conformational changes in glycine tri- and hexapeptide. *European Physical Journal D, 39*, 23–34.
31. Yakubovich, A., Solov'yov, I., Solov'yov, A., & Greiner, W. (2006). Conformations of glycine polypeptides. *Khimicheskaya Fizika (Chemical Physics) 25*, 11–23 (in Russian).
32. Solov'yov, I., Yakubovich, A., Solov'yov, A., & Greiner, W. (2006). Ab initio study of alanine polypeptide chain twisting.*Physical Review E, 73*, 021916–(1–10)
33. Solov'yov, I., Yakubovich, A., Solov'yov, A., & Greiner, W. (2006). Potential energy surface for alanine polypeptide chains. *Journal of Experimental and Theoretical Physics, 102*, 314–326.
34. Yakubovich A., Solov'yov I., Solov'yov A., & Greiner W. (2007). Ab initio theory of helix![CDATA[↔]] coil phase transition. *The European Physical Journal D, 46*, 215–225.
35. Bax, A. (2003). Weak alignment offers new nmr opportunities to study protein structure and dynamics. *Protein Science, 12*, 1–16.
36. Voet, D., & Voet, J. (2004). *Biochemistry*. USA: Wiley.

37. Sheik, S., Sundararajan, P., Hussain, A., & Sekar, K. (2002). Ramachandran plot on the web. *Bioinformatics*. 18:1548–1549.
38. Grimme, S. (2004). Accurate description of van der waals complexes by density functional theory including empirical corrections. *Journal of Computational Chemistry, 25*, 1463–1473.
39. Freindorf, M., & Gao, J. (1996). Optimization of the lennard-jones parameters for a combined ab initio quantum mechanical and molecular mechanical potential using the 3–21g basis set. *Journal of Computational Chemistry, 17*, 386–395.
40. Freindorf, M., Shao, Y., Furlani, T., & Kong, J. (2005). Lennard-Jones parameters for the combined qm/mm method using the b3lyp/6-31+g*/amber potential. *Journal of Computational Chemistry, 26*, 1270–1278.

Chapter 4
Partition Function of a Polypeptide

4.1 Introduction

To study thermodynamic properties of the system one needs to investigate its potential energy surface with respect to all degrees of freedom. There is a number of different methods for calculating the energy of many-body systems. The most accurate approaches are based on solving the Schrödinger equation. These approaches are usually referred to as ab initio methods since they involve a minimum number of assumptions about the system.

For complex molecular systems ab initio calculations require significant computer power. Depending on the method, the computational cost of such calculations grows as N^2 or even N^8 [1], where N is the number of particles in the system. The size of molecular system, which can be described using ab initio methods is therefore limited, and such methods can hardly be used for the description of large biological molecules or systems. This chapter is based on the results published in [2–4].

4.2 Molecular Mechanics Potential

For the description of macromolecular systems, such as polypeptides and proteins, efficient model approaches are necessary. One of the most common tools for the description of macromolecules is based on the so-called molecular mechanics potential, which reads as

$$U = \sum_{i=1}^{N_b} k_i^b (r_i - r_i^0)^2 + \sum_{i=1}^{N_a} k_i^a (\theta_i - \theta_i^0)^2 + \sum_{i=1}^{N_d} k_i^d \left[1 + \cos(n_i \phi_i + \delta_i)\right]$$

$$+ \sum_{i=1}^{N_{id}} k_i^{id} (S_i - S_i^0)^2 + \sum_{\substack{i,j=1 \\ i<j}}^{N} 4\epsilon_{ij} \left[\left(\frac{\sigma_{ij}}{r_{ij}}\right)^{12} - \left(\frac{\sigma_{ij}}{r_{ij}}\right)^{6} \right] + \sum_{\substack{i,j=1 \\ i<j}}^{N} \frac{q_i q_j}{r_{ij}}.$$

$$(4.1)$$

A. V. Yakubovich, *Theory of Phase Transitions in Polypeptides and Proteins*,
Springer Theses, DOI: 10.1007/978-3-642-22592-5_4,
© Springer-Verlag Berlin Heidelberg 2011

Here the first four terms describe the potential energy with respect to variation of distances, angles, dihedral angles and improper dihedral angles between two, three and four neighboring atoms respectively. The last two terms describe the Van der Waals and Coulomb interaction respectively. The summation in the first term goes over all topologically defined bonds in the system, in the second over all topologically defined angles, and in the third over all topologically defined dihedral angles and in the fourth over all topologically defined improper dihedral angles. The total number of bonds, angles, dihedral angles and improper dihedral angles are N_b, N_a, N_d and N_{id} respectively. N is the total number of atoms in the system. k_i^b, k_i^a, k_i^d and k_i^{id} in (4.1) are the stiffness parameters of the corresponding energy terms. r_i^0, θ_i^0 and S_i^0 are the equilibrium values of bonds, angles and improper dihedral angles. n_i and δ_i are the number of possible stable torsion conformations and the initial torsion phase. ϵ_{ij}, σ_{ij} and q_i are the Van der Waals parameters and the charges of atoms in the system.

Parameters k_i^b, k_i^a, k_i^d, k_i^{id}, r_i^0, θ_i^0, S_i^0, n_i, δ_i, ϵ_{ij}, σ_{ij} and q_i are derived from experimental measurements of crystallographic structures, infrared spectra or on the basis of quantum mechanical calculations for small systems (see [5–7] and references therein). The independent variables in (4.1) are r_i, θ_i, ϕ_i and S_i.

Note, that the terms corresponding to the variations of distances, angles and improper dihedral angles in (4.1) describe the motion of the molecule within the harmonic approximation which is reasonable only at low temperatures. The potential energy corresponding to torsion degrees of freedom is usually assumed to be periodic (see 4.1) because several stable conformations of the molecule with respect to these degrees of freedom are possible [5–7]. The torsion degrees of freedom are also referred as the twisting degrees of freedom as discussed in Chap. 3 of the thesis. The most important twisting degrees of freedom for the description of a helix-coil transition in polypeptides are the twisting degrees of freedom along the backbone of the polypeptide [8, 9] (see Sect. 3.2).

4.3 Hamiltonial of a Polypeptide Chain

A Hamiltonian function of a polypeptide chain is constructed as a sum of the potential, kinetic and vibrational energy terms. For a polypeptide chain in a particular conformational state j consisting of n amino acids and N atoms one obtains:

$$H_j = \frac{\mathbf{P}^2}{2M} + \frac{1}{2}\left(I_1^{(j)}\Omega_1^2 + I_2^{(j)}\Omega_2^2 + I_3^{(j)}\Omega_3^2\right) + \sum_{i=1}^{3N-6}\frac{p_i^2}{2m_i} + U(\{x\}), \qquad (4.2)$$

where $\mathbf{P}$, M, $I_{1,2,3}^{(j)}$, $\Omega_{1,2,3}$, are the momentum of the whole polypeptide, its mass, its three main momenta of inertia, and its rotational frequencies. p_i, x_i and m_i are the momentum, the coordinate and the generalized mass describing the motion of the system along the i-th degree of freedom. $U(\{x\})$ is the potential energy of the system, being the function of all atomic coordinates in the system.

One can group all degrees of freedom in a polypeptide in the two classes: "stiff" and "soft" degrees of freedom. The degrees of freedom corresponding to the variation of bond lengths, angles and improper dihedral angles (see Fig. 3.1) are called as "stiff", while degrees of freedom corresponding to the angles φ_i and ψ_i are classified as "soft" degrees of freedom. The "stiff" degrees of freedom can be treated within the harmonic approximation because the energies needed for a noticeable change of the system structure with respect to these degrees of freedom are about several eV which is significantly larger than the characteristic thermal energy of the system at room temperature being on the order of 0.026 eV [5–7, 10, 11].

The Hamiltonian of the polypeptide can be rewritten in terms of the "soft" and "stiff" degrees of freedom. Transforming the set of cartesian coordinates $\{x\}$ to a set of generalized coordinates $\{q\}$, corresponding to the "soft" and "stiff" degrees of freedom one obtains:

$$
H_j = \frac{\mathbf{P}^2}{2M} + \frac{1}{2}\left(I_1^{(j)}\Omega_1^2 + I_2^{(j)}\Omega_2^2 + I_3^{(j)}\Omega_3^2\right) + \sum_{i=1}^{l_s}\sum_{j=l}^{l_s} g_{ij}\frac{p_i^s p_j^s}{2}
$$

$$
+ \sum_{i=1}^{l_s}\sum_{j=l_s+1}^{l_h} g_{ij}\, p_i^s p_j^h + \sum_{i=l_s+1}^{l_h}\sum_{j=l_s+1}^{l_h} g_{ij}\frac{p_i^h p_j^h}{2} + U(\{q^s\}, \{q^h\}), \quad (4.3)
$$

where q^s and q^h are the generalized coordinates corresponding to the "soft" and "stiff" degrees of freedom, and p^s and p^h are the corresponding generalized momenta. l_s and l_h is the number of the "soft" and "stiff" degrees of freedom in the system, satisfying the relation $3N - 6 = l_s + l_h$. $U(\{q^s\}, \{q^h\})$ in (4.3) is the potential energy of the system as a function of the "soft" and "stiff" degrees of freedom. $1/g_{ij}$ has a meaning of the generalized mass, while g_{ij} is defined as follows:

$$
g_{ij} = \sum_{\lambda=1}^{3N-6} \frac{1}{m_\lambda}\frac{\partial q_i}{\partial x_\lambda}\frac{\partial q_j}{\partial x_\lambda}. \tag{4.4}
$$

Here x_λ and m_λ are the generalized coordinate in the cartesian space and the generalized mass of the system, corresponding to the degree of freedom with index λ. q_i and q_j denote the "soft" or the "stiff" generalized coordinate in the transformed space.

The motion of the system with respect to its "soft" and "hard" degrees of freedom occurs on the different time scales as was discussed in [12]. The typical oscillation frequency corresponding to the "soft" degrees of freedom is on the order of $100\,\mathrm{cm}^{-1}$, while for the "stiff" degrees of freedom it is more than $1{,}000\,\mathrm{cm}^{-1}$ [12]. Thus the motion of the system with respect to the "soft" degrees of freedom is uncoupled from the motion of the system with respect to the "stiff" degrees of freedom. Therefore the fifth term in (4.3), which describes the kinetic energy of the "stiff" motions in the polypeptide can be diagonalized. The corresponding set of coordinates $\{\tilde{q}^s\}$ describes the normal vibration modes in the "stiff" subsystem:

$$H_j = \frac{\mathbf{P}^2}{2M} + \frac{1}{2}\left(I_1^{(j)}\Omega_1^2 + I_2^{(j)}\Omega_2^2 + I_3^{(j)}\Omega_3^2\right) + \sum_{i=1}^{l_h}\left(\frac{(\tilde{p}_i^h)^2}{2\mu_i^h} + \frac{\mu_i^h\omega_i^2(\tilde{q}_i^h)^2}{2}\right)$$

$$+ \sum_{i=1}^{l_s}\sum_{j=1}^{l_s} g_{ij}\frac{p_i^s p_j^s}{2} + U(\{\chi\}) + U(\{\varphi,\psi\}). \tag{4.5}$$

Here ω_i and μ_i^h are the frequency of the ith "stiff" normal vibrational mode and the corresponding generalized mass. Note, that the fourth term in (4.3) vanishes if the "soft" and the "stiff" degrees of freedom are uncoupled. The last two terms in (4.5) describe the potential energy of the system in respect to the "soft" degrees of freedom. For every amino acid there are at least two "soft" degrees of freedom, corresponding to the angles φ_i and ψ_i (see Fig. 3.1). Some additional "soft" degrees of freedom involve the rotation of the side radicals in amino acids. A typical example is the angle χ_i, which describes the twisting of the side chain radical along the $C_i^\alpha - C_i^\beta$ bond (see Fig. 3.1). The angle χ_i is defined as the dihedral angle between the planes formed by the atoms $(C_i' - C_i^\alpha - C_i^\beta)$ and by the bonds $C_i^\alpha - C_i^\beta$ and $C_i^\beta - H_{i1}^\beta$. Note, that the notations χ, φ and ψ are used for the simplicity and for the further explanation of the theory. The set of these dihedral angles builds up the set of "soft" degrees of freedom of the polypeptide: $\{q^s\} \equiv \{\chi,\varphi,\psi\}$.

Generalized masses $1/g_{ij}$ depend on the choice of the generalized coordinates in the system. However this dependence can be neglected if the system is considered in the vicinity of its equilibrium state. In this case the motion of the polypeptide with respect to the "soft" degrees of freedom can be considered as the motion of the system of coupled nonlinear oscillators. In the vicinity of the system's equilibrium state the generalized mass can be written as:

$$\frac{1}{g_{ij}} = \frac{1}{g_{ij}\left(\{q_{i0}^s\}\right)} + \sum_{k=1}^{l_s}\left.\frac{\partial\left(1/g_{ij}\right)}{\partial q_k^s}\right|_{q_k^s=q_{k0}^s}\left(q_k^s - q_{k0}^s\right) + o\left(q_k^s - q_{k0}^s\right), \tag{4.6}$$

where q_{k0}^s denotes the value of the kth "soft" degree of freedom at the equilibrium position. The second term in (4.6) describes the dependence of the generalized mass on coordinates and can be neglected if the system is in the vicinity of its equilibrium. All the information about the nonlinearity of the oscillations is contained in the potential energy functions $U(\{\chi\})$ and $U(\{\varphi,\psi\})$ in (4.5).

The validity of the coordinate-independent mass approximation was also discussed in Ref. [12]. The accounting for the coordinate dependence of the generalized masses, g_{ij}, is not performed in the thesis and left further investigation.

4.4 Construction of the Partition Function

The partition function of the polypeptide is constructed within the framework of classical mechanics due to the large masses of the molecules and high temperatures of the conformational transitions. However the presented formalism can be easily generalized for the quantum mechanical description of the system.

All thermodynamic properties of a system are determined by its partition function, which can be expressed via the system's Hamiltonian in the following form [13]:

$$\mathbb{Z} = \int \exp\left(-\frac{H}{kT}\right) d\Gamma, \tag{4.7}$$

where H is the Hamiltonian of the system, k and T are the Boltzmann constant and the temperature respectively and $d\Gamma$ is an element of the phase space. Substituting (4.5) into (4.7) one obtains an expression for the partition function of a polypeptide in a particular conformational state j. Thus, the partition function of the system can be factored as follows:

$$\mathbb{Z} = \frac{1}{(2\pi\hbar)^{3N}} Z_1 \cdot Z_2 \cdot Z_3 \cdot Z_4 \cdot Z_5, \tag{4.8}$$

where

$$Z_1 = \int \exp\left(\frac{1}{kT}\left[-\frac{\mathbf{P}^2}{2M} - \left(\frac{\mathcal{M}_1^2}{2I_1^{(j)}} + \frac{\mathcal{M}_2^2}{2I_2^{(j)}} + \frac{\mathcal{M}_3^2}{2I_3^{(j)}}\right)\right]\right) \times d^3P \cdot d^3Q \cdot d^3\mathcal{M} \cdot d^3\Phi$$
$$= 64\pi^5 V_j M^{3/2}\sqrt{I_j^{(1)} I_j^{(2)} I_j^{(3)}}(kT)^3 \tag{4.9}$$

$$Z_2 = \int \exp\left(-\frac{1}{kT}\sum_{i=1}^{l_h}\left(\frac{(\tilde{p}_i^h)^2}{2\mu_i^h} + \frac{\mu_i^h \omega_i^2 (\tilde{q}_i^h)^2}{2}\right)\right) d^{l_h}\tilde{p}^h \cdot d^{l_h}\tilde{q}^h = \frac{(2\pi kT)^{l_h}}{\prod_{i=1}^{l_h}\omega_i}, \tag{4.10}$$

$$Z_3 = \int \exp\left(-\frac{1}{kT}\sum_{i=1}^{l_s}\frac{(\tilde{p}_i^s)^2}{2\mu_i^s}\right) d^{l_s}\tilde{p}^s = \sqrt{(2\pi kT)}^{l_s}\prod_{i=1}^{l_s}\sqrt{\mu_i^s}, \tag{4.11}$$

$$Z_4 = \int \exp\left(-\frac{U(\{\tilde{\chi}\})}{kT}\right) d^{l_\chi}\tilde{\chi}^s, \tag{4.12}$$

$$Z_5 = \frac{1}{(2\pi\hbar)^{(l_\varphi+l_\psi)/2}} \int \exp\left(-\frac{U(\{\tilde{\varphi},\tilde{\psi}\})}{kT}\right) d^{l_\varphi}\tilde{\varphi}^s \cdot d^{l_\psi}\tilde{\psi}^s. \tag{4.13}$$

Z_1, (4.9), describes the contribution to the partition function originating from the motion of the polypeptide as a rigid body. Here V_j is the specific volume of the polypeptide in conformational state j and $\mathcal{M}$ is the angular momenta of the polypeptide. Z_2, (4.10), accounts for the "stiff" degrees of freedom in the polypeptide. Z_3, (4.11), describes the contribution of the kinetic energy of the "soft" degrees of freedom to the partition function. Z_4, (4.12), and Z_5, (4.13), describe the contribution of the potential energy of the "soft" degrees of freedom to the partition function. Integrating over the phase space in (4.9–4.13) is performed over generalized coordinates and momentum space.

For the derivation of (4.11–4.13) the diagonalization of the quadratic form of the generalized momenta corresponding to the "soft" degrees of freedom in (4.5) is performed and made a transformation $q_i^s \to \tilde{q}_i^s$, $p_i^s \to \tilde{p}_i^s$. In (4.11), μ_i^s is the generalized mass of the i-th "soft" normal vibration mode, being related to g_{ij} in (4.4). $\tilde{\chi}$, $\tilde{\varphi}$ and $\tilde{\psi}$ in (4.12 and 4.13) denote the "soft" twisting degrees of freedom, which have been transformed accordingly. $\tilde{q}_i^s$ and $\tilde{p}_i^s$ are canonical conjugated coordinates. l_χ, l_φ and l_ψ in (4.12 and 4.13) is the number of the χ, φ and ψ degrees of freedom in the system. Note, that $l_s = l_\chi + l_\varphi + l_\psi$.

Integrals in (4.9–4.11) can be evaluated analytically, while for the integration over the angles χ, φ and ψ in (4.12 and 4.13) the knowledge of the exact potential energy surface of the polypeptide is necessary. However the potential energy of the polypeptide corresponding to the twisting degrees of freedom χ does not depend on the conformation of the polypeptide in case of neutral non-polar radicals in simple amino acids (i.e. alanine, glycine) [12]. Thus, the twisting degrees of freedom corresponding to the variations of angles χ have a minor influence on the α-helix$\leftrightarrow$ random coil phase transition. The potential energy of the polypeptide in respect to these degrees of freedom is well described by the following function, as follows from the molecular mechanics potential (4.1):

$$U(\chi_i) = k_{\chi_i}\left[1 + \cos\left(3\chi_i\right)\right], \tag{4.14}$$

where k_{χ_i} is the stiffness parameter of the potential. Since $k_{\chi_i} = k_\chi$, substituting (4.14) into (4.12) and integrating over 2π one obtains:

$$Z_4 = \left[2\pi \exp\left(-\frac{k_\chi}{kT}\right) I_0\left(\frac{k_\chi}{kT}\right)\right]^{l_\chi} = (2\pi)^{l_\chi} B(kT), \tag{4.15}$$

where $I_0(x)$ is the the modified Bessel function of the first kind, and $B(kT) = \left[\exp\left(-\frac{k_\chi}{kT}\right) I_0\left(\frac{k_\chi}{kT}\right)\right]^{l_\chi}$.

Substituting Z_1–Z_5 into (4.8) one obtains the expression for the partition function of a polypeptide in a particular conformational state j:

$$
z_j = \left[\frac{V_j \cdot M^{3/2} \cdot \sqrt{I_j^{(1)} I_j^{(2)} I_j^{(3)}} \, \prod_{i=1}^{l_s} \sqrt{\mu_i^s}}{(2\pi)^{\frac{l_s}{2} - l_x} \pi \, \hbar^{3N} \prod_{i=1}^{l_h} \omega_i} \right] B(kT) \cdot (kT)^{3N-3-\frac{l_s}{2}}
$$

$$
\times \int_{-\pi}^{\pi} \cdots \int_{-\pi}^{\pi} e^{-\frac{U(\{\varphi,\psi\})}{kT}} \, d\varphi_1 \ldots d\varphi_n \, d\psi_1 \ldots d\psi_n
$$

$$
= A_j \cdot B(kT) \cdot (kT)^{3N-3-\frac{l_s}{2}}
$$

$$
\times \int_{-\pi}^{\pi} \cdots \int_{-\pi}^{\pi} e^{-\frac{U(\{\varphi,\psi\})}{kT}} \, d\varphi_1 \ldots d\varphi_n \, d\psi_1 \ldots d\psi_n, \tag{4.16}
$$

A_j denotes the factor in the square brackets. Note, that generalized masses μ_i^h are reduced during the integration and do not enter into the expression of the partition function.

Since a polypeptide exist in different conformational states, one needs to sum over the contributions of all possible conformations Z_j in order to calculate the complete partition function of the polypeptide. For an ensemble of $\mathcal{N}$ noninteracting polypeptides the partition function reads as

$$
\mathbb{Z} = \left(\sum_{j=1}^{\xi} Z_j \right)^{\mathcal{N}} = \left(B(kT) \cdot (kT)^{3N-3-\frac{l_s}{2}} \sum_{j=1}^{\xi} A_j \right.
$$

$$
\left. \times \int_{-\pi}^{\pi} \cdots \int_{-\pi}^{\pi} e^{-\frac{U(\{\varphi,\psi\})}{kT}} \, d\varphi_1 \ldots d\varphi_n \, d\psi_1 \ldots d\psi_n \right)^{\mathcal{N}}, \tag{4.17}
$$

where Z_j is defined in (4.16) and ξ is the total number of possible conformations in a polypeptide. Equation 4.17 has been derived with a minimum number of assumptions about the system. It is general, however, its use for a particular molecular systems is not so straightforward. Expression (4.17) can be further simplified, if one makes additional assumptions about the structure of the system.

For the sake of simplicity, further equations are written for only one polypeptide instead of $\mathcal{N}$. Generalization for the case of $\mathcal{N}$ statistically independent polypeptides can always be done according to (4.17).

One can expect that the factors A_j in (4.17) depend on the chosen conformation of the polypeptide. However, due to the fact that the values of specific volumes, momenta of inertia and frequencies of normal vibration modes of the polypeptide in different conformations are expected to be close [14], the values of A_j in all these conformations can be considered as equal, at least in the zero order approximation. Thus $A_j \equiv A$.

The amino acids can be treated as statistically independent in any conformation of the polypeptide. This fact is not obvious and it was not systematically investigated so far. The statistical independence of small neutral non-polar amino acids (alanine, glycine, etc) in a polypeptide was studied in [15] with the use of time-correlation functions between different amino acids. In Chap. 5, following chapter of the thesis, this question is addressed for alanine polypeptides and determined the degree to which amino acids in the polypeptide can be treated as statistically independent.

With the assumptions made, the partition function of polypeptide reduces to:

$$\mathbb{Z} = A \cdot B(kT) \cdot (kT)^{3N-3-\frac{l_\xi}{2}} \sum_{j=1}^{\xi} \prod_{i=1}^{n} \int_{-\pi}^{\pi} \int_{-\pi}^{\pi} \exp\left(-\frac{\epsilon_i^{(j)}(\varphi, \psi)}{kT}\right) d\varphi d\psi, \quad (4.18)$$

where $\epsilon_i^{(j)}(\varphi, \psi)$ is the potential energy of ith amino acid in the polypeptide, being in one of its ξ conformations denoted with j. The potential energy of the amino acid is calculated as a function of its twisting degrees of freedom φ and ψ.

In (4.18) the partition function is summed over all conformations of the polypeptide. However, in the case of the α-helix to random coil transition of the polypeptide, the summation over the polypeptide conformations has to be performed only over the conformations involved in the transition.

Note that (4.18) is rather general and can be used for the description of the folding process in proteins. Indeed, the partition function in (4.18) is determined by the potential energy surfaces of amino acid in the native state of a protein and in the random coil conformation. The potential energy surfaces can be calculated on the basis of ab initio DFT, combined with molecular mechanics theories as demonstrated in Chap. 5 of the thesis.

Further simplifications of the partition function (4.18) for polypeptide consisting of the identical amino acids can be achieved if one assumes that each amino acid in the polypeptide can occupy two states only, below referred as the *bounded* and *unbounded states*. The amino acid is considered to be in the bounded state when it forms one hydrogen bond with the neighboring amino acids. In the unbounded state amino acids do not have hydrogen bonds. When the α-helix is formed, all amino acids are in the bounded state, while in the case of random coil all amino acids occupy the unbounded states.

All possible conformations of the polypeptide experiencing the α-helix$\leftrightarrow$ random coil phase transition can be divided in four different groups:

1. completely folded state of the polypeptide (α-helix), in which all the amino acids occupy bounded states.
2. partially folded states of the polypeptide (phase co-existence), in which the core of λ amino acids of the polypeptide occupy bounded states, and $n - \lambda$ boundary amino acids are in unbounded states.
3. completely unfolded state of a polypeptide (random coil), in which all the amino acids are in unbounded states
4. phase mixing, in which two or more fragments of a polypeptide are in an α-helix state, while the amino acids between the fragments are in the random coil state.

With the assumptions outlined above and assuming the polypeptide to consist of n identical amino acids the partition function (4.18) of the system can be rewritten as follows:

$$\mathbb{Z} = A \cdot B(kT) \cdot (kT)^{3N-3-\frac{l_s}{2}} \left[\beta Z_b^{n-1} Z_u + \beta \sum_{i=1}^{n-4} (i+1) Z_b^{n-i-1} Z_u^{i+1} + Z_u^n \right.$$

$$\left. + \sum_{i=2}^{(n-3)/2} \beta^i \sum_{k=i}^{n-i-3} \frac{(k-1)!(n-k-3)!}{i!(i-1)!(k-i)!(n-k-i-3)!} Z_b^{k+3i} Z_u^{n-k-3i} \right]$$

$$(4.19)$$

Here the first and the third terms in the square brackets describe the partition function of the polypeptide in the α-helix and in the random coil phases respectively, while the second term in the square brackets accounts for situation of the phase co-existence. The summation in the second term in (4.19) is performed up to $n-4$, because the shortest α-helix consists of four amino acids. The last term in the square brackets accounts for the polypeptide conformations in which a number of amino acids being in the helix conformation are separated by amino acids being in the random coil conformation. The first summation in this term goes over the separated helical fragments of the polypeptide, while the second summation goes over individual amino acids in the corresponding fragment. Polypeptide conformations with two or more helical fragments are energetically unfavorable. This fact is discussed in Chap. 5. As shown in the following chapter, the contribution to the partition function represented by the fourth term in the square brackets in (4.19) is significantly small when compared to the first three terms, for polypeptides containing less than 100 of amino acids. Therefore, it can be omitted in the construction of the partition function. Z_b and Z_u are the contributions to the partition function from a single amino acid being in the bounded or unbounded states respectively, they read as:

$$Z_b = \int_{-\pi}^{\pi} \int_{-\pi}^{\pi} \exp\left(-\frac{\epsilon^{(b)}(\varphi, \psi)}{kT} \right) d\varphi d\psi \qquad (4.20)$$

$$Z_u = \int_{-\pi}^{\pi} \int_{-\pi}^{\pi} \exp\left(-\frac{\epsilon^{(u)}(\varphi, \psi)}{kT} \right) d\varphi d\psi \qquad (4.21)$$

$$\beta = \left(\int_{-\pi}^{\pi} \int_{-\pi}^{\pi} \exp\left(-\frac{\epsilon^{(b)}(\varphi, \psi) + \epsilon^{(u)}(\varphi, \psi)}{kT} \right) d\varphi d\psi \right)^3, \qquad (4.22)$$

where $\epsilon^{(b)}(\varphi, \psi)$ and $\epsilon^{(u)}(\varphi, \psi)$ are the potential energies of a single amino acid being in the bounded or in the unbounded states respectively calculated versus the

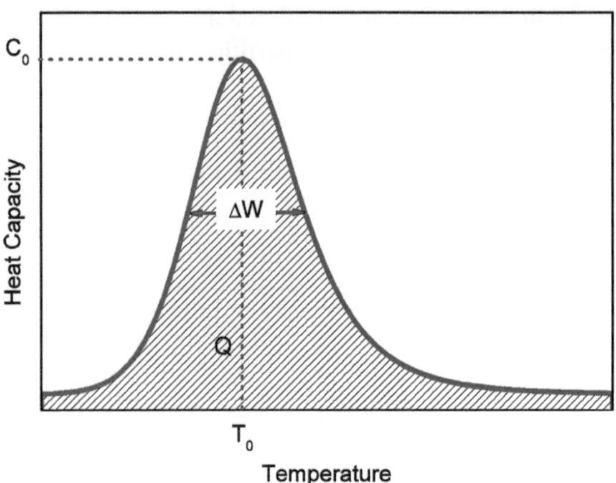

Fig. 4.1 Temperature dependence of the heat capacity for a system experiencing a phase transition. The figure is adopted from [4]

twisting degrees of freedom φ and ψ. β is a factor accounting for the entropy loss of the helix initiation. Substituting (4.20–4.22) into (4.19) one obtains the final expression for the partition function of a polypeptide undergoing an α-helix$\leftrightarrow$ random coil phase transition. This result can be used for the evaluation of all thermodynamical characteristics of the system.

$\epsilon^{(b)}(\varphi, \psi)$ and $\epsilon^{(u)}(\varphi, \psi)$ determine the partition function of a polypeptide. These quantities can be calculated on the basis of ab initio DFT, combined with molecular mechanics theories as demonstrated in Chap. 3 of the thesis.

4.5 Thermodynamical Characteristics of a Polypeptide Chain

The first order phase transition is characterized by an abrupt change of the internal energy of the system with respect to its temperature. In the first order phase transition the system either absorbs or releases a fixed amount of energy while heat capacity as a function of temperature has a sharp peak [13, 16] (see Fig. 4.1).

The peak in the heat capacity is characterized by the transition temperature T_0, the maximal value of the heat capacity C_0, the temperature range of the phase transition ΔW and the specific heat Q, which is also referred as the latent heat of the phase transition (see Fig. 4.1).

All these quantities can be calculated if the dependence of the heat capacity on temperature is known. The temperature dependence of the heat capacity is defined by the partition function as follows [13]:

$$C(T) = kT \frac{\partial^2 T \ln \mathbb{Z}}{\partial T^2}. \tag{4.23}$$

The characteristics of the phase transition are determined by the following equations:

$$\left. \frac{dC(T)}{dT} \right|_{T=T_0} = 0 \tag{4.24}$$

$$C_0 = C(T_0) \tag{4.25}$$

$$C(T_0 \pm \Delta W) = \frac{C_0}{2} \tag{4.26}$$

$$Q = \int_0^\infty C(T)\mathrm{d}T. \tag{4.27}$$

Unfortunately it is not possible to obtain analytical expressions for T_0, C_0, ΔW and Q with partition function defined in (4.19) because the integrals in (4.20) and (4.21) can not be treated analytically. However, the qualitative behavior of these quantities can be understood if one assumes that all conformational states of a polypeptide in a certain phase have the same energy. This model is usually referred to in literature as the two-energy-level model [17] and it turns out to be very useful for the qualitative analysis of the phase transitions in polypeptide chains. If one considers the phase transition between two such phases, the partition function can then be constructed as follows:

$$\mathbb{Z} \approx \mathbb{Z}_0 \left[1 + A \frac{\eta_2}{\eta_1} e^{-\frac{\Delta E}{kT}} \right], \tag{4.28}$$

where $\mathbb{Z}_0$ is the partition function of the system in the first phase, $\Delta E = E_2 - E_1$ is the energy difference between the states of the polypeptide in two different phases, η_1 and η_2 are the numbers of isomeric states of the polypeptide in the first and in the second phases respectively. They can also be considered as the population of the two phases. $A = A_2/A_1$ is the coefficient depending on masses, specific volumes, normal vibration modes frequencies and momenta of inertia of the polypeptide in the two phases. Substituting (4.28) into (4.23) one obtains the expression for the heat capacity in the framework of the two-energy-level model:

$$C(T) = \frac{A \frac{\eta_2}{\eta_1} \Delta E^2 e^{-\left(\frac{\Delta E}{kT}\right)}}{kT^2 \left(1 + A \frac{\eta_2}{\eta_1} e^{-\left(\frac{\Delta E}{kT}\right)} \right)^2}. \tag{4.29}$$

Substituting (4.29) into (4.24–4.27) and solving them one obtains the expressions for T_0, C_0, ΔW and Q, which read as:

$$T_0 \approx \frac{\Delta E}{k \ln \left(A \frac{\eta_2}{\eta_1} \right)} = \frac{\Delta E}{\Delta S}, \tag{4.30}$$

$$C_0 \approx \frac{k}{4} \left[\ln \left(A \frac{\eta_2}{\eta_1} \right) \right]^2 = \frac{\Delta S^2}{4k}, \tag{4.31}$$

$$\Delta W \approx \sqrt{\frac{64 \ln 2}{\pi}} \frac{\Delta E}{k \left[\ln \left(A \frac{\eta_2}{\eta_1} \right) \right]^2} = \sqrt{\frac{64 \ln 2}{\pi}} \frac{k \Delta E}{\Delta S^2} \tag{4.32}$$

$$Q = \int C(T) \mathrm{d}T = \Delta E. \tag{4.33}$$

Here $\Delta S = k \ln A\eta_2 - k \ln \eta_1$ is the entropy change in the system. ΔS and ΔE are the major thermodynamical parameters in the considered problem, since they determine the behavior of the phase transition characteristics. From (4.30–4.32) follows, that $T_0 \sim \frac{\Delta E}{\Delta S}$, $C_0 \sim \Delta S^2$, $Q \sim \Delta E$ and $\Delta W \sim \frac{\Delta E}{\Delta S^2}$.

The numerical calculation and analysis of various thermodynamical characteristics such as the latent heat or the heat capacity is done in the following chapter of the thesis.

References

1. Foresman, J. B., & leen Frisch, A. (1996). *Exploring chemistry with electronic structure methods*. Pittsburgh, PA: Gaussian Inc.
2. Yakubovich, A., Solov'yov, I., Solov'yov, A., & Greiner, W. (2006). Phase transition in polypeptides: a step towards the understanding of protein folding. *European Physical Journal D, 40*, 363–367.
3. Yakubovich, A., Solov'yov, I., Solov'yov, A., & Greiner, W. (2007). Ab initio description of phase transitions in finite bio-nano-systems. *Europhysics News, 38*, 10.
4. Yakubovich, A., Solov'yov, I., Solov'yov, A., & Greiner, W. (2007). Ab initio theory of helix↔ coil phase transition. *European Physical Journal D, 46*, 215–225.
5. MacKerell, A., Bashford, D., Bellott, M., Dunbrack, R., Evanseck, J., & Field, M., et al. (1998). All-atom empirical potential for molecular modeling and dynamics studies of proteins. *The Journal of Physical Chemistry B, 102*, 3586–3616.
6. Scott, W., & van Gunsteren, W. (1995). The GROMOS software package for biomolecular simulations. In E. Clementi and G. Corongiu (Eds.), *Methods and techniques in computational chemistry: METECC-95* (pp. 397–434). Cagliari, Italy: STEF.
7. Cornell, W., Cieplak, P., Bayly, C., Gould, I., Merz, K., & Ferguson, D., et al. (1995). A second generation force field for the simulation of proteins, nucleic acids, and organic molecules. *Journal of the American Chemical Society, 117*, 5179–5197.
8. He, S., & Scheraga, H. A. (1998). Macromolecular conformational dynamics in torsion angle space. *Journal of Chemical Physics, 108*, 271–286.

9. He, S., & Scheraga, H. A. (1998). Brownian dynamics simulations of protein folding. *Journal of Chemical Physics, 108*, 287–300.

10. Solov'yov, I., Yakubovich, A., Solov'yov, A., & Greiner, W. (2006). On the fragmentation of biomolecules: fragmentation of alanine dipeptide along the polypeptide chain. *Journal of Experimental and Theoretical Physics, 103*, 463–471.

11. Solov'yov, I., Yakubovich, A., Solov'yov, A., & Greiner, W. (2006). Potential energy surface for alanine polypeptide chains. *Journal of Experimental and Theoretical Physics, 102*, 314–326.

12. Go, N., & Scheraga, H. A. (1969). Analysis of the contribution of internal vibrations to the statistical weights of equilibrium conformations of macromolecules. *Journal of Chemical Physics, 51*, 4751–4767.

13. Landau, L., & Lifshitz, E. (1959). *Statistical physics*. London-Paris: Pergamon Press.

14. Krimm, S., & Bandekar, J. (1980). Vibrational analysis of peptides, polypeptides, and proteins. v. normal vibrations of β-turns. *Biopolymers, 19*, 1–29.

15. Rubin, A. (2004). *Biophysics: theoretical biophysics*. Nauka: Moscow University Press.

16. Finkelstein, A., & Ptitsyn, O. (2002). *Protein physics. A course of lectures*. Oxford: Elsevier Books.

17. Prabhu, N. V., & Sharp, K. A. (2005). Heat capacity in proteins. *The Annual Review of Physical Chemistry, 56*, 521–548.

Chapter 5
Phase Transitions in Polypeptides

5.1 Introduction

In the previous chapter a novel and general theoretical method for the description of phase transitions in finite complex molecular systems was introduced. In particular, it was demonstrated that for polypeptide chains, i.e., chains of amino acids, one can identify specific twisting degrees of freedom responsible for the folding dynamics of these amino acid chains. In other words, these degrees of freedom characterize the transition from a chain in a random coil state into that in an α-helix structure and vice versa.

The essential domains of the potential energy surface (PES) of polypeptides with respect to these twisting degrees of freedom have been calculated and thoroughly analyzed on the basis ab initio density functional theory (DFT). In Chap. 4 of the thesis, it was shown that knowing the PES, one can construct a partition function of a polypeptide chain, from which it is then possible to extract all essential thermodynamical variables and properties, such as the heat capacity, phase transition temperature, free energy, etc.

In this chapter, the above introduced formalism is further explored and applied to a detailed analysis of the α-helix $\leftrightarrow$ random coil phase transition in alanine polypeptides of different lengths. This system was chosen because it has been widely investigated both theoretically [1–17] and experimentally [18–21] during the last five decades (for review see, e.g. [22–25]) and thus is a perfect system for testing a novel theoretical approach.

The theoretical studies of the helix–coil transition in polypeptides have been performed both with the use of statistical mechanics methods [1–6, 13–17, 25] and of molecular dynamics (MD) [9–13]. Previous attempts to describe the helix–coil transition in polypeptide chains using the principles of statistical mechanics were based on the models suggested in 1960s [1–4]. These models were based on the construction of the polypeptide partition function depending on several parameters and were widely used in Refs. [13–17, 22, 24, 25] for the description of the helix–coil transition in polypeptides.

A. V. Yakubovich, *Theory of Phase Transitions in Polypeptides and Proteins*,
Springer Theses, DOI: 10.1007/978-3-642-22592-5_5,
© Springer-Verlag Berlin Heidelberg 2011

For a comprehensive overview of the relevant work see recent reviews [23–25] and the book [22].

Experimentally, extensive studies of the helix–coil transition in polypeptides have been conducted [18–21]. In Ref. [18], the enthalpy change of an α-helix to random coil transition for the Ac-Y(AEAAKA)$_8$F-NH$_2$ peptide in water was determined calorimetrically. The dependence of the heat capacity of the polypeptide on temperature was measured using differential scanning calorimetry. In Refs. [19, 20], UV resonance Raman spectroscopy was performed on the MABA-[A]$_5$-[AAARA]$_3$-ANH$_2$ peptide. Using circular dichroism methods, the dependence of helicity on temperature was measured. In Ref. [21], the kinetics of the helix–coil transition of the 21-residue alanine polypeptide was investigated by means of infrared spectroscopy.

In this chapter, the PES of polyalanines of different lengths were calculated with respect to their twisting degrees of freedom. This was done within the framework of classical molecular mechanics. However, to scrutinize the accuracy of these calculations, was performed a comparison of the resultant molecular mechanics potential energy landscapes with those obtained using ab initio density functional theory (DFT). The comparison was only performed for alanine tripeptide and hexapeptide, since for larger polypeptides, the DFT calculation becomes increasingly computationally demanding. Hence for these larger systems, only molecular mechanics simulations have been used.

The calculated PES was then used to construct a parameter-free partition function of the polypeptide using the statistical method outlined in Chap. 4. This partition function was then used to derive various thermodynamical characteristics of alanine polypeptides as a function of temperature and polypeptide length. The temperature dependence of the heat capacity, latent heat and helicity of alanine polypeptides consisting of 21, 30, 40, 50 and 100 amino acids was calculated and analyzed. A correspondence between the presented ab initio method with the results of the semiempirical approach of Zimm and Bragg [1] was also established and analyzed. Thus, the key parameters of the Zimm–Bragg theory were determined using the presented approach.

Finally, the heat capacity, latent heat and helicity of alanine polypeptides were calculated using molecular dynamics and the obtained results were compared with those using the statistical approach. Comparison between the two methods allows one to establish the accuracy of the statistical method for relatively small molecular systems, and lets us gauge the feasibility of extending the description to larger molecular objects for which it is especially essential in those cases where MD simulations are hardly possible due to computational limitations.

The chapter is organized as follows. In Sect. 6.2.1 the final expressions obtained within the formalism described in Chap. 4 are presented and the basic equations and the set of parameters, which have been used in MD calculations, are introduced. The results of this chapter are published in [26]. In Sect. 5.3 the results of computer simulations obtained with the use of developed theoretical method are presented and discussed. Then the results are compared with results of MD simulations. The work presented in Sect. 5.3 is published in [27].

5.2 Molecular Dynamics Simulations

MD is an approach which is widely used for the study of dynamics of macromolecular systems [28–30]. Within the framework of MD, one has to solve the equations of motion for all particles in the system interacting via a given potential. Since the technique of MD is well known and described in numerous textbooks [29, 31, 32], here are presented only the basic equations and ideas underlying this method.

MD simulations usually imply the numerical solution of the Langevin equation [32–34]:

$$m_i \mathbf{a_i} = m_i \ddot{\mathbf{r_i}} = -\frac{\partial U(\mathbf{R})}{\partial \mathbf{r_i}} - \beta_i \mathbf{v_i} + \mathbf{\Omega}(t). \tag{5.1}$$

Here m_i, $\mathbf{r_i}$, $\mathbf{v_i}$ and $\mathbf{a_i}$ are the mass, radius vector, velocity and acceleration of the ith atom. $U(\mathbf{R})$ is the potential energy of the system. The second term on the right hand side describes the viscous force which is proportional to the particle velocity. The proportionality constant $\beta_i = m_i \gamma$, where γ is the damping coefficient. The third term is the random force term originating from collisions of the molecule with atoms in the medium. In the MD formalism the system of Langevin equations for all particles is being integrated over time.

In this chapter the CHARMM27 force field [35] is used to describe the interactions between atoms. This is a common empirical force field for treating polypeptides, proteins and lipids [28, 35–37]. The set of the parameters used in the simulations can be found in Refs. [28, 29, 31, 32]. All simulations were performed using the NAMD molecular dynamics program [29], while visualization of the results was done with VMD [38].

The polypeptide was considered in the NVT canonical ensemble and the heat capacity of the system was calculated using two different approaches. The first approach is based on the calculation of the heat capacity from the derivative of the average energy of the system:

$$C_v = \left.\frac{\partial \langle E \rangle}{\partial T}\right|_{V=\text{const}}, \tag{5.2}$$

where T is the temperature of the system and $\langle E \rangle$ is the time-averaged value of the polypeptide energy. Knowing the value of $\langle E \rangle$ the heat capacity of the polypeptide can be calculated. However, since the MD simulations are performed for a limited number of different temperatures and for the finite time the direct numeric differentiation of $\langle E \rangle$ can lead to large numerical artifacts in the heat capacity. For a better analysis the values of $\langle E \rangle$ can either be interpolated and smoothed using a standard numerical procedure (see e.g. Ref. [39]).

$\langle E \rangle$ can also be calculated from the partition function of the system as follows:

$$\langle E \rangle = \frac{\sum_i E_i e^{-\frac{E_i}{kT}}}{\sum_i e^{-\frac{E_i}{kT}}}, \tag{5.3}$$

where E_i is the energy of the ith state, and k is the Bolzmann factor. The summation in (5.3) is performed over all accessible states of the system. Substituting (5.3) into (5.2) and performing differentiation one obtains:

$$C_v = \frac{\sum_i E_i^2 e^{-\frac{E_i}{kT}} \sum_i e^{-\frac{E_i}{kT}} - \sum_i E_i e^{-\frac{E_i}{kT}} \sum_i E_i e^{-\frac{E_i}{kT}}}{kT^2 \left(\sum_i e^{-\frac{E_i}{kT}}\right)^2} = \frac{\langle E^2 \rangle - \langle E \rangle^2}{kT^2}, \quad (5.4)$$

where $\langle E^2 \rangle$ is the average value of the energy square, which is defined as:

$$\langle E^2 \rangle = \frac{\sum_i E_i^2 e^{-\frac{E_i}{kT}}}{\sum_i e^{-\frac{E_i}{kT}}}. \quad (5.5)$$

Equation (5.4) shows that the heat capacity of a polypeptide can be defined from the energy fluctuations in the system. In Sect. 5.4 the discussed methods are applied for the study of heat capacity in alanine, valine and leucine polypeptides.

5.3 α-Helix $\leftrightarrow$ Random Coil Phase Transition in Polyalanine

In this section are presented the results of calculations obtained using the statistical mechanics approach and those from the MD simulations. In Sect. 5.3.1 is discussed the accuracy of Molecular Mechanics force field as applied to alanine polypeptides. In Sect. 5.3.2 are presented the PESs for different amino acids in alanine polypeptide calculated versus the twisting degrees of freedom φ and ψ (see Fig. 3.1). In Sect. 5.4.2, the statistical mechanics approach is used for the description of the α-helix $\leftrightarrow$ random coil phase transition. Here, the results of the statistical mechanics approach are compared to those obtained from MD simulations. In Sect. 5.3.7 the statistical independence of amino acids in the polypeptide is discussed.

5.3.1 Accuracy of the Molecular Mechanics Potential

The PES of alanine polypeptides was calculated using the CHARMM27 force field [35] that has been parameterized for the description of proteins, in particular polypeptides, and lipids. Nevertheless, the level of its accuracy when applied to alanine polypeptides cannot be taken for granted and has to be investigated. Therefore, the PESs for alanine tri- and hexapeptide calculated using the CHARMM27 force field are compared with those calculated using ab initio density functional theory (DFT). In the DFT approach, the PES of alanine tri- and hexapeptides were calculated as a function of the twisting degrees of freedom, φ and ψ in the central amino acid of the polypeptide(see Fig. 3.1). All other degrees of freedom were frozen.

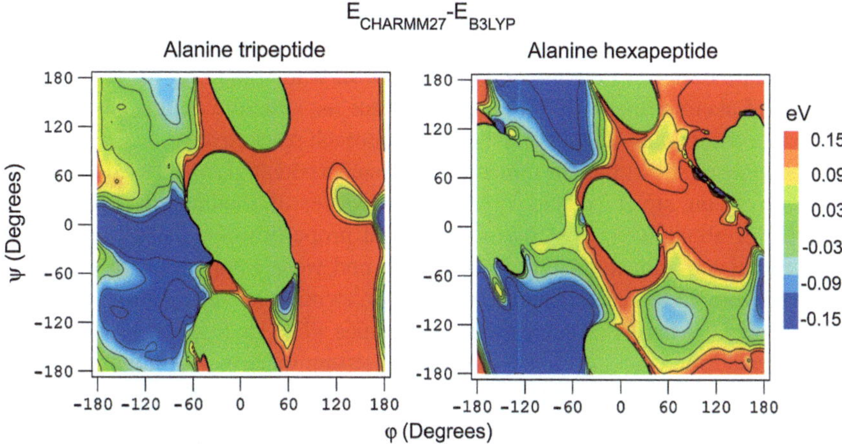

Fig. 5.1 Difference between the PESs calculated with the CHARMM27 force field and with the B3LYP functional (see Fig. 3.4) for the alanine tripeptide (*left*) and the alanine hexapeptide (*right*). The relative energies are given in eV. The equipotential lines are shown for the energies -0.10, -0.050, 0.05 and 0.1 eV. The figure is adopted from [26]

To establish the accuracy of the CHARMM27 force field, were calculated the PESs of alanine polypeptides in its β-sheet conformation. The geometry of alanine tri- and hexapeptide used in the calculations are shown in Fig. 3.2a, b respectively. The ab initio calculations were performed in Sect. 3.2.2 using B3LYP, Becke's three-parameter gradient-corrected exchange functional [40] with the gradient-corrected correlation functional of Lee et al. [41]. The wave function of all electrons in the system was expanded using a standard basis set B3LYP/6-31G(2d,p). The PESs calculated within the DFT approach have been analyzed in Sect. 3.2.4.

The difference between the PESs calculated with the CHARMM27 force field and with the B3LYP functional is shown in Fig. 5.1 for the alanine tripeptide (left plot) and for the alanine hexapeptide (right plot).

From Fig. 5.1, one can conclude that the energy difference between the PESs calculated with the CHARMM27 force field and with the B3LYP functional is <0.15 eV. To describe the relative deviation of the PESs, the relative error of the two methods is introduced as follows:

$$\eta = \frac{2 \int |E_{B3LYP}(\varphi, \psi) - E_{CHARMM27}(\varphi, \psi)| d\varphi d\psi}{\int |E_{B3LYP}(\varphi, \psi) + E_{CHARMM27}(\varphi, \psi)| d\varphi d\psi} \times 100\%, \qquad (5.6)$$

where $E_{B3LYP}(\varphi, \psi)$ and $E_{CHARMM27}(\varphi, \psi)$ are the potential energies calculated within the DFT and molecular mechanics methods respectively. Calculating η for alanine tri- and hexapeptide, one obtains: $\eta_{3 \times Ala} = 27.6\%$ and $\eta_{6 \times Ala} = 23.4\%$ respectively. These values show that the molecular mechanics approach is reasonable for a qualitative description of the alanine polypeptide. Note however, that the PES obtained for alanine hexapeptide within the molecular mechanics method is closer to the PES calculated within the DFT approach. This occurs because the PESs $E_{CHARMM27}(\varphi, \psi)$ and $E_{B3LYP}(\varphi, \psi)$ of alanine hexapeptide were calculated for the structure optimized within the DFT approach, while the PESs $E_{CHARMM27}$ and E_{B3LYP} of alanine tripeptide were calculated for the structure optimized within the molecular mechanics method and the DFT approach respectively.

The analysis shows that the molecular mechanics potential can be used to describe qualitatively the structural and dynamical properties of alanine polypeptides with an error of about 20%. The thermodynamical properties of alanine polypeptides were calculated with the use of MD method and were compared with the results obtained from the statistical approach. However, ab initio MD calculations of alanine polypeptides are hardly possible on the time scales when the α-helix $\leftrightarrow$ random coil phase transition occurs, even for systems consisting of only 4–5 amino acids (see discussion in Sect. 3.1). Therefore, the MD simulations for alanine polypeptides were performed using molecular mechanics forcefield. In order to establish the accuracy of the statistical mechanics approach, the PES used for the construction of the partition function was also calculated with the same method.

5.3.2 Potential Energy Surface of Alanine Polypeptide

To construct the partition function (see (4.19)), one needs to calculate the PES of a single amino acid in the bounded, $\epsilon^{(b)}(\varphi, \psi)$, and unbounded, $\epsilon^{(u)}(\varphi, \psi)$, conformations versus the twisting degrees of freedom φ and ψ (see Fig. 3.1). The potential energies of alanine in different conformations determine the Z_b and Z_u contributions to the partition function, defined in (4.20)–(4.21).

The PES of an alanine depends both on the conformation of the polypeptide and on the amino acid index in the chain. The PES for different amino acids of the 21-residue alanine polypeptide calculated as a function of twisting dihedral angles φ and ψ are shown in Fig. 5.2. These surfaces were calculated with the use of the CHARMM27 forcefield for a polypeptide in the α-helix conformation. The PESs (a)–(e) in Fig. 5.2 correspond to the variation of the twisting angles in the second, third, fourth, fifth and tenth amino acids of the polypeptide respectively. Amino acids are numbered starting from the NH_2 terminal of the polypeptide. The PES for the amino acids at boundary is not presented because the angle φ is not defined for it.

On the PES corresponding to the tenth amino acid in the polypeptide (see Fig. 5.2e), one can identify a prominent minimum at $\varphi = -81°$ and $\psi = -71°$. This minimum corresponds to the α-helix conformation of the corresponding amino acid, and energetically, the most favorable amino acid configuration. In the α-helix conformation the tenth amino acid is stabilized by two hydrogen bonds (see Fig. 5.3). With the change of the twisting angles φ and ψ, these hydrogen bonds become broken and the energy of the system increases. The tenth alanine can form hydrogen bonds with the neighboring amino acids only in the α-helix conformation, because all other amino acids in the polypeptide are in this particular conformation. This fact is clearly seen from the corresponding PES Fig. 5.2e, where all local minima

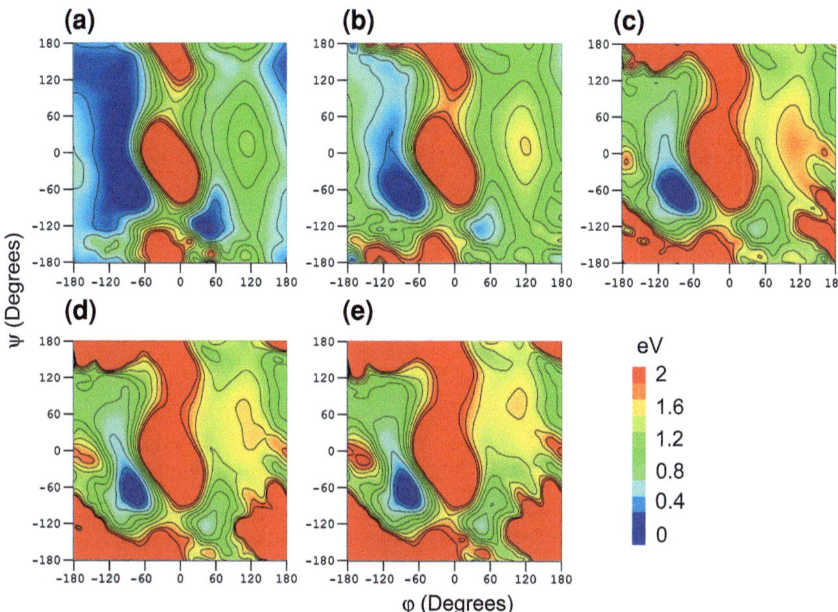

Fig. 5.2 PESs for different amino acids of alanine polypeptide consisting of 21 amino acids calculated as the function of twisting dihedral angles φ and ψ in: **a** second alanine, **b** third alanine, **c** fourth alanine **d** fifth alanine and **e** tenth alanine. Amino acids are numbered starting from the NH_2 terminal of the polypeptide. Energies are given with respect to the lowest energy minimum of the PES in eV. The equipotential lines are shown for the energies 1.8, 1.6, 1.4, 1.2, 1.0, 0.8, 0.6, 0.4 and 0.2 eV. The figure is adopted from [26]

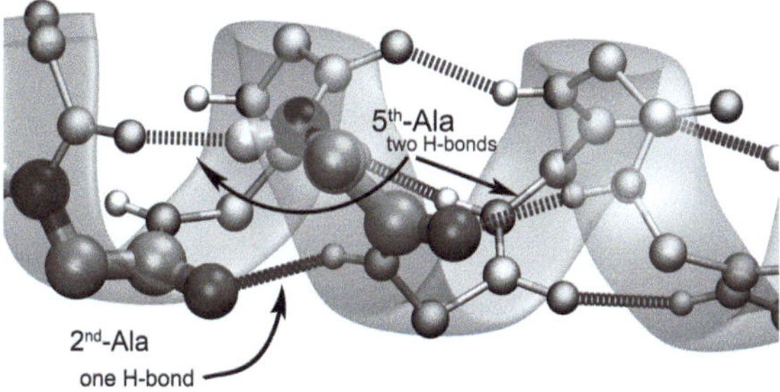

Fig. 5.3 Alanine polypeptide in the α-helix conformation. *Dashed lines* show the hydrogen bonds in the system. Figure shows that the second alanine forms only one hydrogen bond, while the fifth alanine forms two hydrogen bonds with the neighboring amino acids. The figure is adopted from [26]

have energies significantly higher than the energy of the global minima (the energy difference between the global minimum and a local minimum with the closest energy is $\Delta E = 0.736$ eV, which is found at $\varphi = 44°$ and $\psi = -124°$).

The PES depends on the amino acid index in the polypeptide. This fact is clearly seen from Fig. 5.2. The three boundary amino acids in the polypeptide form a single hydrogen bond with their neighbors (see Fig. 5.3) and therefore are more weakly bounded than the amino acids inside the polypeptide. The change in the twisting angles φ and ψ in the corresponding amino acids leads to the breaking of hydrogen bonds, hence increasing the energy of the system. However, the boundary amino acids are more flexible then those inside the polypeptide chain, and therefore their PESs are smoother.

Figure 5.2 shows that the PESs calculated for the fourth, fifth and the tenth amino acids are very close and have minor deviations from each other. Therefore, the PESs for all amino acids in the polypeptide, except the boundary ones can be considered identical.

Each amino acid inside the polypeptide forms two hydrogen bonds. However since these bonds are shared by two amino acids, there is only effectively one hydrogen bond per amino acid (see Fig. 5.3). Therefore, to determine the potential energy surface of a single amino acid in the bounded, $\epsilon^{(b)}(\varphi, \psi)$, and unbounded, $\epsilon^{(u)}(\varphi, \psi)$, conformations is used the potential energy surface calculated for the second amino acid of the alanine polypeptide (see Fig. 5.2a), because only this amino acid forms single hydrogen bond with its neighbors (see Fig. 5.3).

The PES of the second amino acid Fig. 5.2a has a global minimum at $\varphi = -81°$ and $\psi = -66°$, that corresponds to the bounded conformation of the alanine. Therefore the part of the PES in the vicinity of this minimum corresponds to the PES of the bounded state of the polypeptide, $\epsilon^{(b)}(\varphi, \psi)$. The potential energy of the bounded state is determined by the energy of the hydrogen bond, which for an alanine is equal to $E_{HB} = 0.142\,\text{eV}$. This value is obtained from the difference between the energy of the global minimum and the energy of the plateaus at $\varphi \in (-90° \ldots -100°)$ and $\psi \in (0° \ldots 60°)$ (see Fig. 5.2a). Thus, the part of the potential energy surface which has an energy less then E_{HB} corresponds to the bounded state of alanine, while the part with energy greater then E_{HB} corresponds to the unbounded state.

In Fig. 5.4 are presented the potential energy surfaces for alanine in both the bounded (plot a) and unbounded (plot b) conformations. Both PESs were calculated from the PES for the second amino acid in the polypeptide, which is shown in plot (c) of Fig. 5.4.

5.3.3 Internal Energy of Alanine Polypeptide

Knowing the PESs for all amino acids in the polypeptide, one can construct the partition function of the system using (4.19). Plots (a) and (b) in Fig. 5.4 show the dependence of $\epsilon^{(b)}(\varphi, \psi)$ and $\epsilon^{(u)}(\varphi, \psi)$ on the twisting angles φ and ψ, while $\epsilon^{(b)}$ and $\epsilon^{(u)}$ define the contributions of the bounded and unbounded states of the polypeptide to the partition function of the system (see (4.20)–(4.21)). The expressions for Z_b and Z_u are integrated numerically and the partition function of the polypeptide

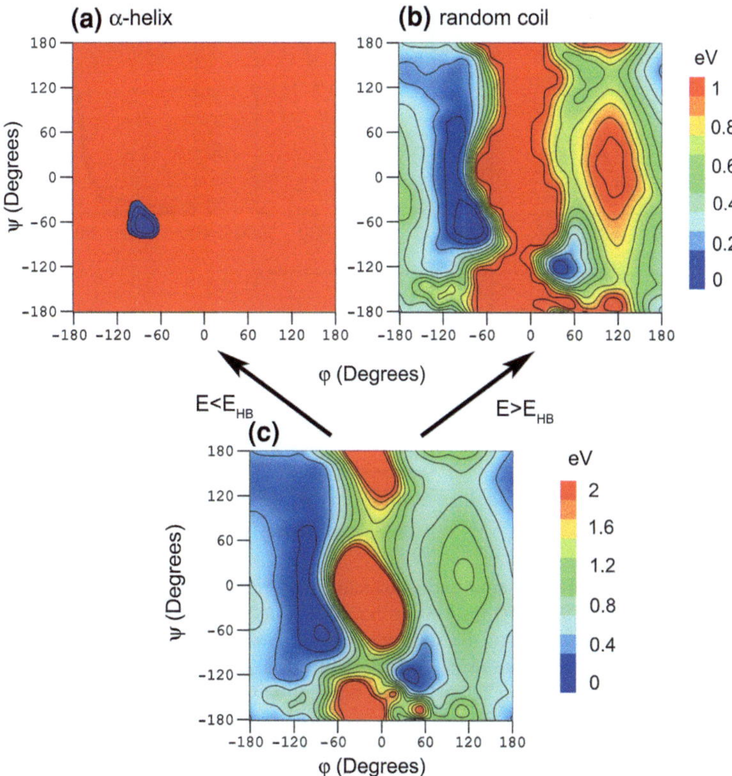

Fig. 5.4 PESs for alanine in α-helix (plot **a**) and random coil conformation (plot **b**). The potential energy surface for the second amino acid of the polypeptide is shown in plot **c** and is used to determine the PESs for alanine in α-helix and random coil conformations. The part of the PES shown in plot **c**, with energy less then E_{HB} corresponds to the α-helix conformation (bounded state) of the alanine, while the part of the potential energy surface with energy greater then E_{HB} corresponds to the random coil conformation (unbounded state). The energies are given in eV. The equipotential lines in plot **a** are shown for the energies 0.05 and 0.1 and 0.15 eV; in plot **b** for the energies 0.1, 0.2, 0.3, 0.4, 0.5, 0.6, 0.7, 0.8 and 0.9 eV; in plot **c** for the energies 1.8, 1.6, 1.4, 1.2, 1.0, 0.8, 0.6, 0.4 and 0.2 eV. The figure is adopted from [26]

is evaluated according to (4.19). The partition function defines all essential thermodynamical characteristics of the system as discussed in Chap. 4.

The first order phase transition is characterized by an abrupt change of the internal energy of the system with respect to its temperature. In the first order phase transition the system either absorbs or releases a fixed amount of energy while the heat capacity as a function of temperature has a pronounced peak [22, 23, 42, 43]. The manifestation of these peculiarities is studied for alanine polypeptide chains of different lengths.

Figure 5.5 shows the dependencies of the internal energy on temperature calculated for alanine polypeptides consisting of 21, 30, 40, 50 and 100 amino acids. The thick solid lines correspond to the results obtained using the statistical approach, while the dots show the results of MD simulations. From Fig. 5.5 it is seen that the

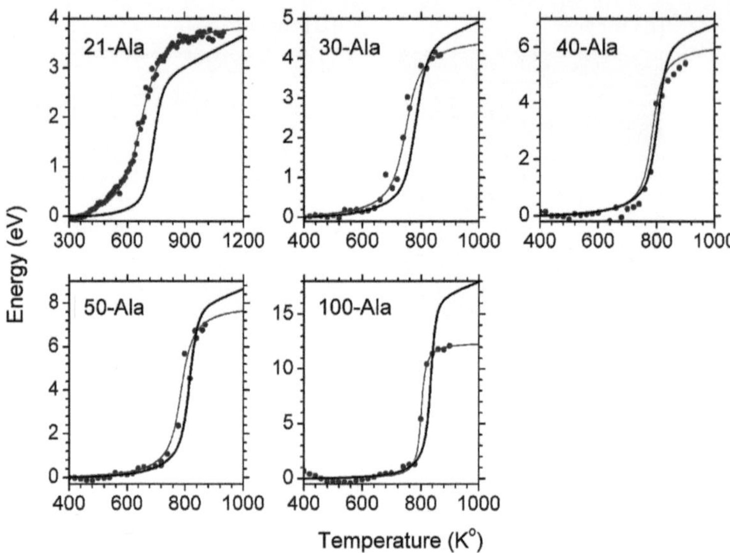

Fig. 5.5 Dependencies of the internal energy on temperature calculated for the alanine polypeptide chains consisting of 21, 30, 40, 50 and 100 amino acids. *Thick solid lines* correspond to the results obtained within the framework of the statistical model. *Dots* correspond to the results of MD simulations, which are fitted using (5.7). The fitting functions are shown with *thin solid lines*. The fitting parameters are compiled in Table 5.1. The figure is adopted from [26]

internal energy of alanine polypeptide rapidly increases in the vicinity of a certain temperature corresponding to the temperature of the first order phase transition. The value of the step-like increase of the internal energy is usually referred as the latent heat of the phase transition denoted as Q. The latent heat is the energy that the system absorbs at the phase transition. Figure 5.5 shows that the latent heat increases with the growth of the polypeptide length. This happens because in the α-helix state, long polypeptides have more hydrogen bonds than short ones and, for the formation of the random coil state, more energy is required.

The characteristic temperature region of the abrupt change in the internal energy (half-wight of the heat capacity peak) characterizes the temperature range of the phase transition. This quantity is denoted as ΔT. With the increase of the polypeptide length the dependence of the internal energy on temperature becomes steeper and ΔT decreases. Therefore, the phase transition in longer polypeptides is more pronounced. In the following subsection is discussed in detail the dependence of ΔT on the polypeptide length.

With the molecular dynamics, one can evaluate the dependence of the total energy of the system on temperature, which is the sum of the potential, kinetic and vibrational energies. Then the heat capacity can be factorized into two terms: one, corresponding to the internal dynamics of the polypeptide and the other, to the potential energy of the polypeptide conformation. The conformation of the polypeptide influences only the term related to the potential energy and the term corresponding to the internal dynamics is assumed to be independent of the polypeptides conformation.

Table 5.1 Parameters used in (5.7) to fit the results of MD simulations

n	E_0	$\Delta E/\pi$	γ	T_0	a
21	11.38 ± 0.24	1.37 ± 0.10	79.4 ± 7.6	670.0 ± 2.0	0.0471 ± 0.0003
30	13.61 ± 0.58	1.50 ± 0.16	37.9 ± 7.3	747.4 ± 3.3	0.0699 ± 0.0008
40	16.80 ± 0.39	1.991 ± 0.083	26.6 ± 2.2	785.7 ± 1.8	0.0939 ± 0.0005
50	19.94 ± 0.79	2.59 ± 0.21	29.4 ± 5.5	786.6 ± 2.9	0.118 ± 0.0010
100	29.95 ± 0.67	4.00 ± 0.16	10.5 ± 2.0	801.1 ± 1.1	0.2437 ± 0.0009

This factorization allows one to distinguish from the total energy the potential energy term corresponding to the structural changes of the polypeptide. The formalism of this factorization is discussed in detail in Sect. 4.4. The energy term corresponding to the internal dynamics of the polypeptide neither influence the phase transition of the system, nor grows linearly with temperature. The term corresponding to the potential energy of the polypeptide conformation has a step-like dependence on temperature that occurs at the temperature of the phase transition. Since the work is focused on the manifestation of the phase transition, linear term is subtracted from the total energy of the system and only its non-linear part is considered. The slope of the linear term was obtained from the dependencies of the total energy on temperature in the range of 300–450° K, which is far beyond the phase transition temperature (see Fig. 5.5). Note that the dependence shown in Fig. 5.5 corresponds only to the non-linear potential energy terms.

The heat capacity of the system is defined as the derivative of the total energy on temperature. However, as seen from Fig. 5.5 the MD data is scattered in the vicinity of a certain expectation line. Therefore, the direct differentiation of the energy obtained within this approach will lead to non-physical fluctuations of the heat capacity. To overcome this difficulty a fitting function for the total energy of the polypeptide is defined as follows:

$$E(T) = E_0 + \frac{\Delta E}{\pi} \arctan \left[\frac{T - T_0}{\gamma} \right] + aT, \tag{5.7}$$

where E_0, ΔE, T_0, γ and a are the fitting parameters. The first and the second terms are related to the potential energy of the polypeptide conformation, while the last term describes the linear increase of the total energy with temperature. The fitting function (5.7) was used for the description of the total energy of polypeptides in earlier papers [12, 44]. The results of fitting are shown in Fig. 5.5 with the thin solid lines. The corresponding fitting parameters are compiled in Table 5.1

Figure 5.5 shows that the results obtained using the MD approach are in a reasonable agreement with the results obtained from the statistical mechanics formalism. The fitting parameter ΔE corresponds to the latent heat of the phase transition, while the temperature width of the phase transition is related to the parameter γ. With the increase of the polypeptides length, the temperature width of the phase transition decreases (see γ in Table 5.1), while the latent heat increases (see ΔE in Table 5.1). These features are correctly reproduced in MD and in the statistical mechanics approach.

Table 5.2 Parameters, characterizing the heat capacity peak in Fig. 5.6 calculated using the statistical approach

n	C_{300} (meV/K)	T_0 (K)	C_0 (eV/K)	ΔT (K)	Q (eV)
21	1.951	740	0.027	90	1.741
30	2.725	780	0.051	75	2.727
40	3.584	805	0.084	55	3.527
50	4.443	815	0.123	50	4.628
100	8.740	835	0.392	29	8.960

Heat capacity at 300 K, C_{300}, the transition temperature T_0, the maximal value of the heat capacity C_0, the temperature range of the phase transition ΔT and the specific heat Q are shown as a function of polypeptide length, n

Furthermore, MD simulations demonstrate that with an increase of the polypeptide length, the temperature of the phase transition shifts towards higher temperatures (see Fig. 5.5). The temperature of the phase transition is described by the fitting parameter T_0 in Table 5.1. Note also, that the increase of the phase transition temperature is reproduced correctly within the framework of the statistical mechanics approach, as seen from Fig. 5.5.

Nonetheless, the results of MD simulations and the results obtained using the statistical mechanics formalism have several discrepancies. As seen from Fig. 5.5 the latent heat of the phase transition for long polypeptides obtained within the framework of the statistical approach is higher than that obtained in MD simulations. This happens because within the statistical mechanics approach, the potential energy of the polypeptide is underestimated. Indeed, long polypeptides (consisting of more than 50 amino acids) tend to form short-living hydrogen bonds in the random coil conformation. These hydrogen bonds lower the potential energy of the polypeptide in the random coil conformation. However, the "dynamic" hydrogen-bonds are neglected in the present formalism of the partition function construction.

Additionally, the discrepancies between the two methods arise due to the limited MD simulation time and to the small number of different temperatures at which the simulations were performed. Indeed, for alanine polypeptide consisting of 100 amino acids 26 simulations were performed, while only 3–5 simulations correspond to the phase transition temperature region (see Fig. 5.5).

5.3.4 Heat Capacity of Alanine Polypeptide

The dependence of the heat capacity on temperature for alanine polypeptides of different lengths is shown in Fig. 5.6. The results obtained using the statistical approach are shown with the thick solid line, while the results of MD simulations are shown with the thin solid line. Since the classical heat capacity is constant at low temperatures, this constant value is subtracted out for a better analysis of the phase transition in the system. The constant contribution to the heat capacity is denoted as C_{300} and it is calculated as the heat capacity value at 300° K. The C_{300} values

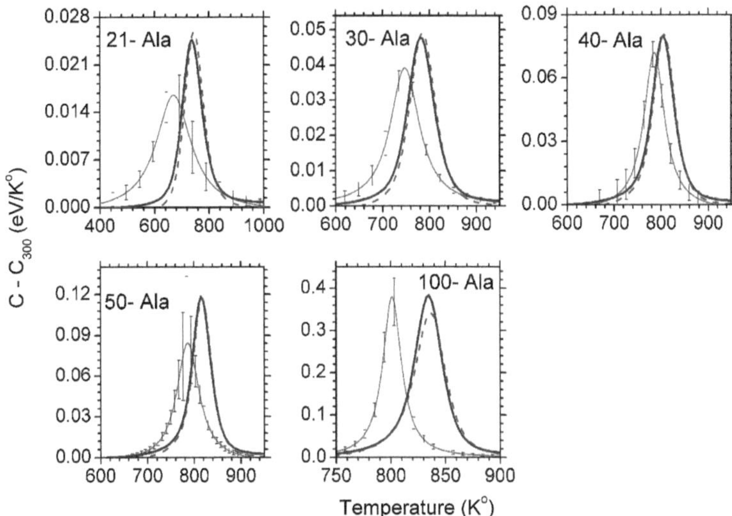

Fig. 5.6 Dependencies of the heat capacity on temperature calculated for the alanine polypeptides consisting of 21, 30, 40, 50 and 100 amino acids. The results obtained using the statistical approach are shown with the *thick solid line*, while the results of MD simulations are shown with the *thin solid line*. *Dashed lines* show the heat capacity as a function of temperature calculated within the framework of the Zimm–Bragg theory [1]. C_{300} denotes the heat capacity at 300° K, which are compiled in Table 5.2. The figure is adopted from [26]

for alanine polypeptides of different length are compiled in the second column of Table 5.2.

As seen from Fig. 5.6, the heat capacity of the system as a function of temperature acquires a sharp maximum at a certain temperature corresponding to the temperature of the phase transition. The peak in the heat capacity is characterized by the transition temperature T_0, the maximal value of the heat capacity C_0, the temperature range of the phase transition ΔT and the latent heat of the phase transition Q. These parameters have been extensively discussed in Sect. 4.5. Within the framework of the two-energy level model describing the first order phase transition, it is shown that:

$$
\begin{aligned}
T_0 &\sim \frac{\Delta E}{\Delta S} = \text{const} \\
C_0 &\sim \Delta S^2 \sim n^2 \\
Q &\sim \Delta E \sim n \\
\Delta T &\sim \frac{\Delta E}{\Delta S^2} \sim \frac{1}{n}.
\end{aligned}
\tag{5.8}
$$

Here ΔE and ΔS are the energy and the entropy changes between the α-helix and the random coil states of the polypeptide, while n is the number of amino acids in the polypeptide. Figure 5.7 shows the dependence of the α-helix $\leftrightarrow$ random coil phase transition characteristics on the length of the alanine polypeptide. The maximal heat

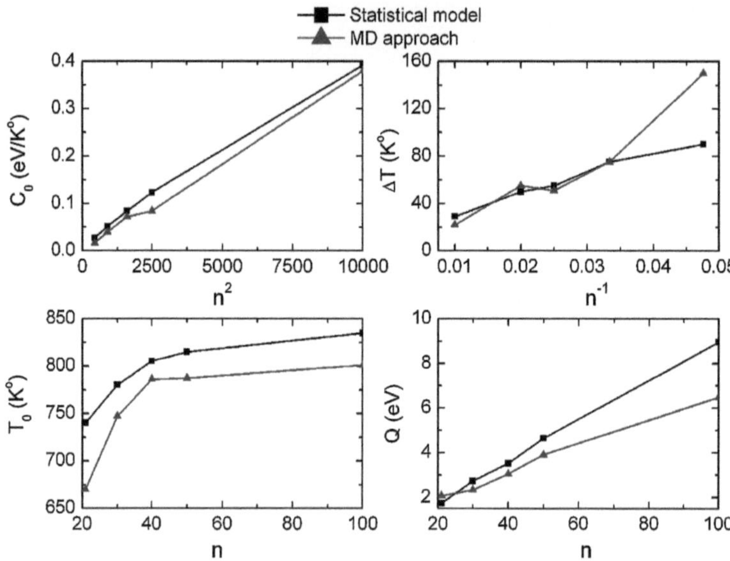

Fig. 5.7 Phase transition parameters C_0, ΔT, T_0 and Q calculated as a function of polypeptide length. *Squares* and *triangles* represent the phase transition parameters calculated using the statistical approach and those obtained from the MD simulations respectively. The figure is adopted from [26]

capacity C_0 and the temperature range of the phase transition ΔT are plotted against the squared number of amino acids (n^2) and the inverse number of amino acids ($\frac{1}{n}$) respectively, while the temperature of the phase transition T_0 and the latent heat of the phase transition Q are plotted against the number of amino acids (n). Squares and triangles represent the phase transition parameters calculated using the statistical approach and those obtained from the MD simulations respectively.

The results obtained within the framework of the statistical model are in a good agreement with the results obtained on the basis of MD simulations. The relative deviation of the phase transition parameters calculated in both methods is on the order of 10% for short polypeptides and 5% for long polypeptides, as follows from Fig. 5.7. However, since the MD simulations are computationally time demanding it is difficult to simulate phase transition in large polypeptides. The difficulties arise due to the large fluctuations which appear in the system at the phase transition temperature and to the large time scale of the phase transition process. The relative error of the phase transition temperature obtained on the basis of MD approach is in the order of 3–5%, while the relative error of the heat capacity is about 30% in the vicinity of the phase transition (see Fig. 5.6).

At present, there are no experiments devoted to the study of phase transition of alanine polypeptides in vacuo, but such experiments are feasible and are already planned.[1] In Ref. [16] the temperature of the α-helix $\leftrightarrow$ random coil phase transition was calculated. Depending on the parameter set, the temperature of the transition ranges from 620 to 650° K for right-handed α-helix, and from 730 to 800 K for a left-handed α-helix.

The heat capacity peak is asymmetric. The heat capacity at higher temperatures, beyond the heat capacity peak, is not zero and forms a plateau (see Fig. 5.6). The plateau is formed due to the conformations of the amino acids with larger energies (see Sect. 3.2.5). At $T = 1000$ K), the difference in the heat capacity of the polypeptide is 7.6×10^{-4}, 1.2×10^{-3}, 1.6×10^{-3}, 2.1×10^{-3} and 4.3×10^{-3} eV/K for the Ala_{21}, Ala_{30}, Ala_{40}, Ala_{50} and Ala_{100} peptides respectively. The magnitude of the plateau increases with the growth of the polypeptide length. This happens because the number of energy levels with high energies rapidly increases for longer polypeptide chains.

5.3.5 Calculation of the Zimm–Bragg Parameters

An alternative theoretical approach for the study of α-helix $\leftrightarrow$ random coil phase transition in polypeptides was introduced by Zimm and Bragg [1]. It is based on the construction of the partition function of a polypeptide involving two parameters s and σ, where s describes the contribution of a bounded amino acid relative to that of an unbounded one, and σ describes the entropy loss caused by the initiation of the α-helix formation.

The Zimm–Bragg theory [1] is semiempirical because it is parameter dependent. The theoretical method described in the preceding chapter of the thesis (Chap. 4) and which is used in the present chapter is different as it does not include any parameters and the construction of the partition function is based solely on the PES of a polypeptide. Therefore, the construction of the partition function as described in the thesis is free of any parameters, and this is what makes it different from the models suggested previously. Assuming that the polypeptide has a single helical region, the partition function derived within the Zimm–Bragg theory, reads as:

$$Q = 1^n + \sigma \sum_{k=1}^{n-3} (n - k - 2)s^k, \tag{5.9}$$

where $n + 1$ is the number amino acids in the polypeptide, s and σ are the parameters of the Zimm–Bragg theory. The partition function, which is used in the present work (4.19) can be rewritten in a similar form:

$$Z = \left[1 + \beta s(T)^3 \sum_{k=1}^{(n-1)-3} (n - k - 3)s(T)^k \right] \xi(T). \tag{5.10}$$

---

[1] Helmut Haberland (private communication).

Here n is the number of amino acids in the polypeptide and the functions $s(T)$ and $\xi(T)$ are defined as:

$$s(T) = \frac{\int_{-\pi}^{\pi} \int_{-\pi}^{\pi} \exp\left(-\frac{\epsilon^{(b)}(\varphi,\psi)}{kT}\right) d\varphi d\psi}{\int_{-\pi}^{\pi} \int_{-\pi}^{\pi} \exp\left(-\frac{\epsilon^{(u)}(\varphi,\psi)}{kT}\right) d\varphi d\psi} \qquad (5.11)$$

$$\xi(T) = \left[\int_{-\pi}^{\pi} \int_{-\pi}^{\pi} \exp\left(-\frac{\epsilon^{(u)}(\varphi, \psi)}{kT}\right) d\varphi d\psi \right]^n , \qquad (5.12)$$

where $\epsilon^{(b)}(\varphi, \psi)$ and $\epsilon^{(u)}(\varphi, \psi)$ are the potential energies of a single amino acid in the bounded and unbounded conformations respectively calculated versus its twisting degrees of freedom φ and ψ. By comparing (5.9) and (5.10), one can evaluate the Zimm–Bragg parameters as:

$$\sigma(T) = \beta(T)s(T)^3, \qquad (5.13)$$

where $\beta(T)$ is defined in (4.22).

The dependence of the Zimm–Bragg parameters s and σ on temperature is shown in Fig. 5.8a, b respectively. The function $-RT \ln(s)$ grows linearly with an increase in temperature, as seen in Fig. 5.8a. The zero of this function corresponds to the temperature of the phase transition in an infinitely long polypeptide. In the present calculation it is $860\,K$ (see black line in Fig. 5.8a). Parameter σ is shown in the logarithmic scale and has a maximum at $T = 560° \,K$. Note, that this maximum does not correspond to the temperature of the phase transition.

The parameters of the Zimm–Bragg theory were considered in earlier papers [13, 16, 45]. In Fig. 5.8a is presented the dependence of parameter s on temperature calculated in [16] (see squares, triangles and stars in Fig. 5.8b) using a matrix approach described in Ref. [3]. The energies of different polypeptide conformations were calculated using the force field described in Ref. [46]. Squares, triangles and stars correspond to three different force field parameter sets used in Ref. [16], which are denoted as sets A, B and C. Figure 5.8a shows that the results obtained in the thesis are closer to the results obtained using the parameter set C. This figure also illustrates that the Zimm–Bragg parameter s depends on the parameter set used. Therefore, the discrepancies between the presented calculation and the calculation performed in Ref. [16] arise due to the utilization of different force fields.

The Zimm–Bragg parameter σ was also calculated in Ref. [16]. However, it was not systematically studied for the broad range of temperatures, and therefore it is not plot in Fig. 5.8b. In Ref. [16] the parameter σ was calculated only for the temperature of the α-helix $\leftrightarrow$ random coil phase transition ranging from 620 to 800° K. In Ref. [16], it was also demonstrated that parameter σ is very sensitive to the force field parameters, being in the range $10^{-9.0}$ to $10^{-3.6}$. In the performed calculation $\sigma = 10^{-3.4}$ at 860 K. The dependence of the parameter σ on the force field parameters was extensively discussed in Ref. [16], where it was demonstrated that this parameter

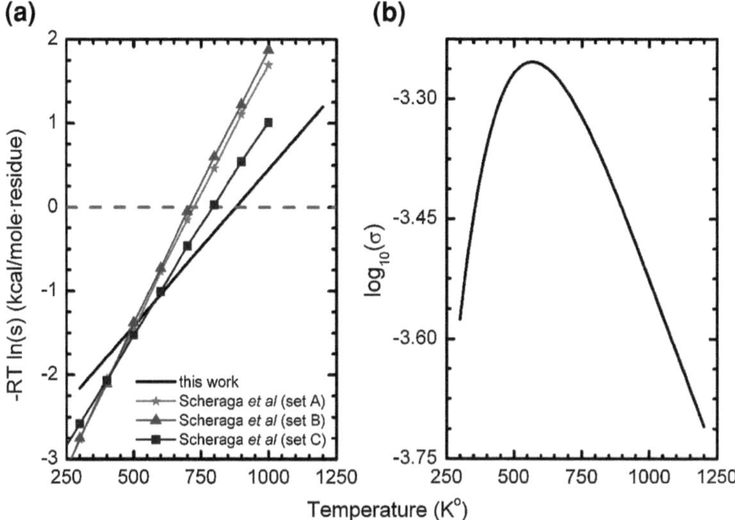

Fig. 5.8 Dependence of the parameters of the Zimm–Bragg theory [1] s (plot **a**) and σ (plot **b**) on temperature. Parameter s describes the contribution to the partition function of a bounded amino acid relative to that of an unbounded one. The parameter σ describes the entropy loss caused by the initiation of the α-helix formation. Parameter s was also calculated in Ref. [16] using three different force fields, shown with *stars*, *triangles* and *squares* in plot **a**. The figure is adopted from [26]

does not have a strong influence on the thermodynamical characteristics of phase transition.

If the parameters s and σ are known, it is possible to construct the partition function of the polypeptide in the form suggested by Zimm and Bragg [1], and on its basis calculate all essential thermodynamic characteristics of the system. The dependence of the heat capacity calculated within the framework of the Zimm–Bragg theory is shown in Fig. 5.6 by dashed lines for polypeptides of different length.

From Fig. 5.6 it is seen that results obtained on the basis of the Zimm–Bragg theory are in a perfect agreement with the results of the statistical approach. The values of the phase transition temperature and of the maximal heat capacity in both cases are close. The comparison shows that the heat capacity obtained within the framework of the Zimm–Bragg model at temperatures beyond the phase transition window is slightly lower than the heat capacity calculated within the framework of the statistical model.

An important difference of the Zimm–Bragg theory from the theory described in the thesis arises due to the accounting for the states of the polypeptide with more than one α-helix fragment. These states are often referred to as multihelical states of the polypeptide. However, their statistical weight in the partition function is suppressed. The suppression arises because of entropy loss in the boundary amino acids of a helical fragment. The boundary amino acids have weaker hydrogen bonds than amino acids in the central part of the α-helix. At the same time the entropy of such amino acids is smaller than the entropy of an amino acids in the coil state. These two factors lead to the decrease of the statistical weight of the multihelical states.

The contribution of the multihelical states to the partition function leads to the broadening of the heat capacity peak while the maximal heat capacity decreases. The multihelical states become important in longer polypeptide chains that consist of more than 100 amino acids. As seen from Fig. 5.6, the maximal heat capacity obtained within the framework of the Zimm–Bragg model for Ala_{100} polypeptide is 10% lower than that obtained using the suggested statistical approach. For alanine polypeptide consisting of less than 50 amino acids the multihelical states of the polypeptide can be neglected as seen from the comparison performed in Fig. 5.6. Omission of the multihelical states significantly simplifies the construction and evaluation of the partition function.

5.3.6 Helicity of Alanine Polypeptides

Helicity is an important characteristic of the polypeptide which can be measured experimentally [18–21]. It describes the fraction of amino acids in the polypeptide that are in the α-helix conformation. With the increase of temperature the fraction of amino acids being in the α-helix conformation decreases due to the α-helix $\leftrightarrow$ random coil phase transition. The helicity of a polypeptide can be defined as follows:

$$f_\alpha = \frac{\sum_{i=0}^{n-4}(i+1)(n-i-1)Z_u^{i+1}Z_b^{n-i-1}}{n\left(Z_u^n + \beta \sum_{i=1}^{n-4}(i+1)Z_u^{n+1}Z_b^{n-i-1} + \beta Z_b^{n-1}Z_u\right)},$$

where n is the number of amino acids in the polypeptide, Z_b, Z_u are the contributions to the partition function from amino acids in the bounded and unbounded states defined in (4.20) and (4.21) respectively. The dependencies of helicity on temperature obtained using the statistical approach for alanine polypeptides of different length are shown in Fig. 5.9.

On the basis of MD simulations, it possible to evaluate the dependence of helicity on temperature. Helicity can be defined as the ratio of amino acids being in the α-helix conformation to the total number of amino acids in the polypeptide, averaged over the MD trajectory. The amino acid is considered to be in the conformation of an α-helix if the angles describing its twisting are within the range of $\varphi \in [-72°; -6°]$ and $\psi \in [0°; -82°]$. This region was chosen from the analysis of angles φ and ψ distribution at 300 K. The helicity for alanine polypeptide consisting of 21 amino acids obtained within the framework of MD approach is shown in the inset to Fig. 5.9. From this plot it is seen that at $T \approx 300$ K, which is far beyond the temperature of the phase transition, the helicity of the Ala_{21} polypeptide is 0.82. The fact that at low temperatures the helicity of the polypeptide obtained within the MD approach is smaller than unity arises due to the difficulty of defining the α-helix state

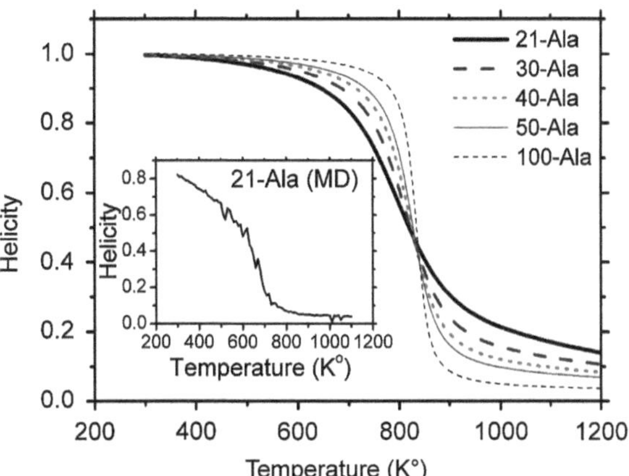

Fig. 5.9 Dependency of the helicity on temperature obtained using the statistical approach for alanine polypeptide chains consisting of 21, 30, 40, 50 and 100 amino acids. The helicity for alanine polypeptide consisting of 21 amino acids obtained within a framework of MD approach is shown in the *inset*. The figure is adopted from [26]

of an amino acid. Thus, the helicity obtained within the MD approach rolls off at lower temperatures compared to the helicity of the polypeptide of the same length obtained using the statistical mechanics approach.

The kink in the helicity curve corresponds to the temperature of the phase transition of the system. As seen from Fig. 5.9, with an increase of the polypeptide length, the helicity curve is becomes steeper as the phase transition is getting sharper. In the limiting case of an infinitely long polypeptide chain, the helicity should behave like a step function. This is yet another feature of a first-order phase transition.

5.3.7 Correlation of Different Amino Acids in the Polypeptide

An important question concerns the statistical independence of amino acids in the polypeptide at different temperatures. The influence of a particular conformation of one amino acids on the PESs of other amino acids is analyzed in this subsection. In Fig. 5.10 are presented the deviations of angles φ and ψ from the twisting angles φ_{10} and ψ_{10} in the 10 th amino acid of alanine polypeptide. These results were obtained on the basis of MD simulations of the Ala_{21} polypeptide at $300°$ K and at $1000°$ K. The deviation of angles φ and ψ is defined as follows:

$$\text{RMSD}(\varphi_i) = \sum_{j=1}^{j<=M} \sqrt{\frac{1}{M}(\varphi_i - \varphi_{10})^2}$$

$$\text{RMSD}(\psi_i) = \sum_{j=1}^{j<=M} \sqrt{\frac{1}{M}(\psi_i - \psi_{10})^2}, \tag{5.14}$$

where i is the amino acid index in the polypeptide and M is the number of MD simulation steps. Note, that the plots shown in Fig. 5.10 do not depend on the reference amino acid (the middle amino acid in the polypeptide was used).

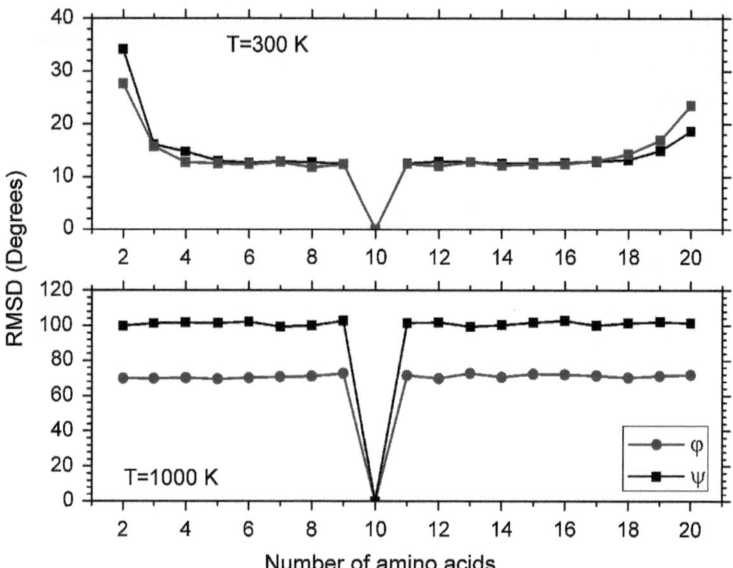

Fig. 5.10 The root mean square deviation of angles φ and ψ calculated with the use of (5.14) for alanine polypeptide consisting of 21 amino acids. The calculations were done in respect to the tenth amino acid of the polypeptide for 300 K (*top plot*) and for 1000 K (*bottom plot*). The figure is adopted from [26]

The top plot in Fig. 5.10 was obtained at 300° K. At this temperature, all amino acids in the polypeptide are in the α-helix conformation, and the deviation of angles φ and ψ is <16° for all amino acids except the boundary ones, where the relative deviation of the angles φ and ψ is 28° and 34° respectively. This happens because, while the boundary amino acids are loosely bounded, the central amino acids in the polypeptide are close to the minimum that corresponds to an α-helix conformation. In the α-helix state, all central amino acids are stabilized by two hydrogen bonds, while the boundary amino acids form only one hydrogen bond.

At 1000° K the polypeptide is, to large extent, found in the random coil phase and therefore becomes more flexible. In the random coil phase, the stabilizing hydrogen bonds are broken, and the deviation of angles φ and ψ significantly increases. This fact is clearly seen in the bottom plot of Fig. 5.10. However at 1000 K, the deviation of angles φ and ψ in the central and in the boundary amino acids is almost the same, confirming the assumption that in the random coil phase, short alanine polypeptides do not build hydrogen bonds.

Another important fact which is worth mentioning is that in the random coil phase (and in the central part of the α-helix), the deviation of angles φ and ψ does not depend on the distance between amino acids in the polypeptide chain. For instance, the deviation between angles in the 10th and in the 11th amino acid is almost the same as the deviation between angles in the 10th and in the 17th amino acid. This fact allows one to conclude that in a certain phase of the polypeptide (α-helix or random coil), amino acids can be treated as statistically independent.

5.4 Phase Transitions in Polypeptides: Analysis of Energy Fluctuations

MD approach (an alternative to using statistical physics) has been widely used during the last decades for studying structural transitions in polypeptides. Full atomistic molecular dynamics [9–13, 47, 48] and Monte-Carlo based techniques [12, 13, 49, 50, 51] were used for studying alanine tripeptide [9], alanine pentapeptide [10], alanine 13- and 15-peptide [47, 52], alanine 21-peptide [11, 13], mixed alanine-rich peptide [49] and Ala_x peptide (with $x = 21, 30, 40, 50, 100$) (see Sect. 5.4.2). The molecular dynamics simulations were carried out within the framework of classical mechanics with an empirical Hamiltonian usually referred as a forcefield.

MD simulations allow one to calculate thermodynamical characteristics of a system. Thus, the dependence of the heat capacity on temperature is of primary importance because it can be measured experimentally and reveals important features of a phase transition (i.e. the phase transition temperature, temperature range for the phase transition, the maximum heat capacity). The heat capacity of a system can be calculated as the derivative of the system's internal energy or derived from the energy fluctuations. Both methods have been used [12, 47–51], but no comparison between them have been made so far. It is not clear which one is more accurate and thus preferable. The present section is devoted to the discussion of this question and elucidation of the limitations of both theoretical approaches by considering phase transitions in polypeptides. For this purpose is studied the helix$\leftrightarrow$ random coil transition in alanine, valine and leucine polypeptides consisting of 30 amino acids and calculated the heat capacity as a function of temperature. The discrepancies between the results obtained with the use of the two different theoretical methods are analyzed.

It is also shown that in the course of the helix$\leftrightarrow$ random coil phase transition the polypeptide chain can experience several sequential structural changes leading to the emergence of additional peaks in the temperature dependence of the heat capacity. It is illustrated on the example of the Val_{30} polypeptide, where two phase transitions can be observed. The origin of both transitions is elucidated by demonstration that the main transition has all the features of the phase transition leading to the destruction of the polypeptide secondary structure, while another one is associated with the order$\leftrightarrow$ disorder transition in side chain radicals.

5.4.1 Fluctuations of Internal Energy and Heat Capacity

In this section are presented the results of molecular dynamics simulations performed for alanine, valine and leucine polypeptides consisting of 30 amino acids. At certain temperatures all polypeptides undergo the helix$\leftrightarrow$ random coil transition, which can be understood as a first order phase transition in finite systems.

The structure of the alanine, valine and leucine amino acids is shown in Fig. 5.11. The three amino acids are neutral and non-polar, and differ from each other by the

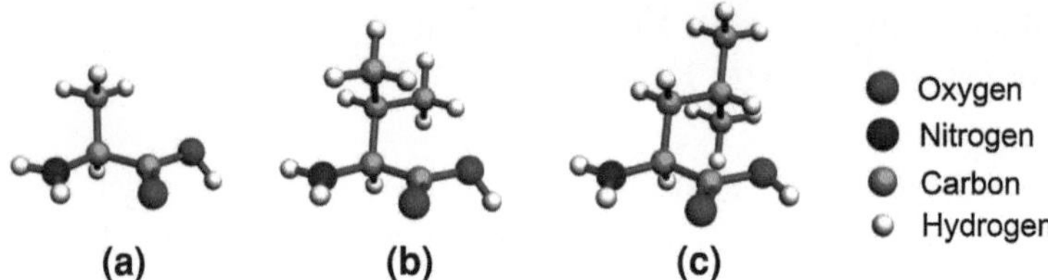

Fig. 5.11 Structure of alanine (**a**), valine (**b**) and leucine (**c**) amino acids. The figure is adopted from [27]

side chain radical, which is CH_3-, C_3H_7- and C_4H_9- in the case of alanine, valine and leucine respectively.

In spite of similarities of the alanine, valine and leucine, the polypeptides consisting of these amino acids have different conformations in their ground states. Therefore in vacuo alanine polypeptides undergo the α-helix $\leftrightarrow$ random coil transition, while valine and leucine ones undergo the π-helix $\leftrightarrow$ random coil transition. These processes are discussed in Sects. 5.4.2, 5.4.3 and 5.4.4 respectively.

5.4.2 α-Helix $\leftrightarrow$ Random Coil Transition in Alanine Polypeptide

For the study of helix–coil transition in polypeptides one needs to calculate the transition energy which is defined as:

$$E_{trans} = E_{total} - 2E_{kinetic}, \tag{5.15}$$

where E_{total} is the total energy of the polypeptide and $E_{kinetic}$ is its kinetic energy. The transition energy is a convenient characteristic of structural transitions of a molecular systems as it behaves nonlinearly in the vicinity of the transition temperature. If all the interactions within the system are harmonic (the potential energy is proportional to the squared displacement from the equilibrium position) the average (over time) potential energy is equal to the average kinetic energy of the system. For harmonic systems the total (internal) energy is proportional to the temperature. Therefore, transition energy characterizes the deviation of the system from harmonicity, which is a signature of structural changes in the system, such as α-helix $\leftrightarrow$ random coil transition.

Figure 5.12a shows the dependence of transition energy on temperature calculated for the Ala_{30} polypeptide. The squares correspond to the results obtained with MD simulations, while the solid line shows the fitting of MD simulation results by a cubic B-spline. Figure 5.12a shows that the transition energy of the alanine polypeptide rapidly increases in the vicinity of certain temperature corresponding to the temperature of the first order like phase transition. The value of the step-like growth of the transition energy is the latent heat of the phase transition. It is equal to the energy absorbed by the system in the course of the phase transition.

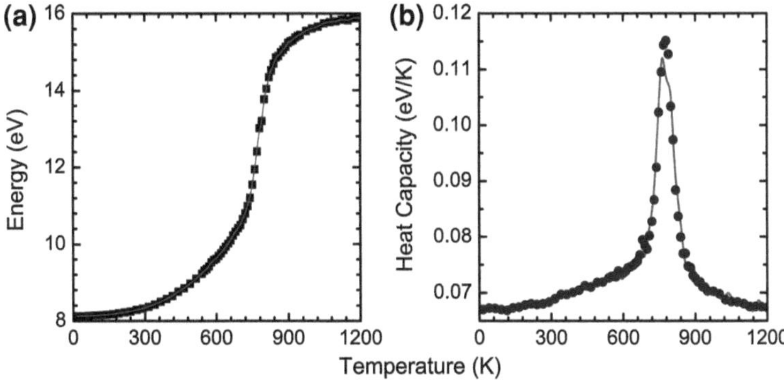

Fig. 5.12 Dependence of transition energy, (5.15), on temperature calculated for the Ala$_{30}$ polypeptide (**a**) and corresponding dependence of the heat capacity on temperature (**b**). In **a**: the *squares* show the results obtained from MD simulations and the *solid line* shows the cubic B-spline interpolation. In **b**: the *dots* show dependence of the heat capacity calculated via the analysis of polypeptide's energy fluctuations; the *solid line* corresponds to the derivative of the interpolating function of the transition energy on temperature. The figure is adopted from [27]

The heat capacity of the system can be obtained either as a derivative of the internal energy with respect to temperature or from analysis of the energy fluctuations (see Sect. 5.2). In Fig. 5.12b is shown dependence of the heat capacity on temperature calculated using both approaches. As seen from Fig. 5.12b, the heat capacity of the system as a function of temperature acquires a sharp maximum at a certain temperature corresponding to the temperature of the phase transition.

The α-helix $\leftrightarrow$ random coil transition in alanine polypeptides has been widely studied, because alanine polypeptide is a relatively simple system comparing to other polypeptides (see e.g. Refs. [9, 17, 18, 20] and references therein). According to the CHARMM27 forcefield the temperature of the α-helix $\leftrightarrow$ random coil transition in Ala$_{30}$ polypeptide is 780 K (see Fig. 5.12b). During 500 ns (see the details of performed MD simulations in Sect. 5.4.5.) the polypeptide continuously changes its conformation between different states with a characteristic transition time of $\sim$10 ns for the α-helix $\rightarrow$ random coil and backward transitions. This results in approximately 50 α-helix $\leftrightarrow$ random coil transitions during the performed simulation leading to a good statistics of energy fluctuations.

5.4.3 π-Helix $\leftrightarrow$ Random Coil Transition in Valine Polypeptide

The α-helix conformation is not the global energy minimum for valine polypeptide in vacuo, since π-helix conformation has lower energy, according to the CHARMM27 [35] forcefield. Therefore in this subsection is studied the π-helix $\leftrightarrow$ random coil transition in valine polypeptide consisting of 30 amino acids. In the π-helix conformation the N–H group of an amino acid forms a hydrogen bond with the C=O group

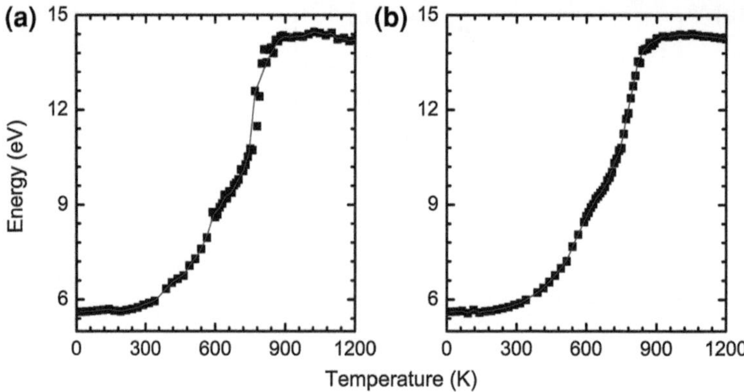

Fig. 5.13 Dependence of transition energy, (5.15), on temperature calculated for the Val_{30} polypeptide. The *squares* show the results obtained from MD simulations and the *thin line* shows the interpolating function. Part **b** shows the results of the total simulation (see Sect. 5.4.5), while part **a** represents the results obtained in a simulation being of 16 times shorter duration. The figure is adopted from [27]

of another amino acid being placed five residues away, while in the α-helix conformation this hydrogen bond is formed between the amino acids being four residues away from each other.

In Fig. 5.13b is shown the dependence of transition energy, (5.15), on temperature calculated for the Val_{30} polypeptide (see the details of performed MD simulations in Sect. 5.4.5). The simulation time should be chosen to be long enough in order to ensure that the heat capacity does not depend on the simulation time. In Fig. 5.13a are shown the results obtained in a simulation, for which the simulation time was 16 times shorter than in the case Fig. 5.13b. From the results of MD simulations presented in Fig. 5.13a, b it is possible to calculate the heat capacity of the system and compare the results of different methods of its calculation (fluctuation of the energy in the system and differentiation of the energy).

The dependence of heat capacity on temperature is shown in Fig. 5.14. Figure 5.14b shows that for the Val_{30} polypeptide there are two well pronounced peaks, while in Fig. 5.14a (shorter simulation), the first peak is not clearly seen. In Fig. 5.14, the dots show the temperature dependence of the heat capacity obtained from the analysis of polypeptide's energy fluctuations, while the solid line corresponds to the derivative of the interpolating function of the transition energy on temperature. The oscillations of the heat capacity in Fig. 5.14a allow one to estimate the accuracy of the methods as being both on the order of 20%.

The energy fluctuation approach is more general than the method based on differentiation of the internal energy on temperature. It does not depend on the number of data points (simulations at different temperatures) and allows one to determine the absolute values of the heat capacity. Indeed, for the Val_{30} polypeptide, 5–9 simulations are sufficient to reproduce both peaks in the heat capacity on temperature

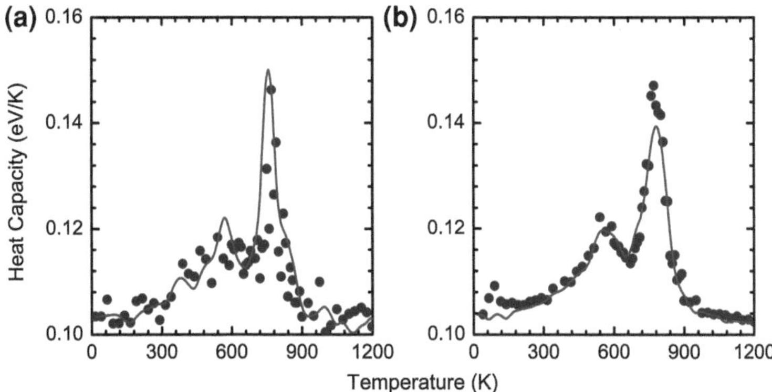

Fig. 5.14 Dependence of the heat capacity on temperature calculated for the Val$_{30}$ polypeptide. Plots **a** and **b** correspond to the energy dependencies shown in Fig. 5.13a, b respectively. The *dots* show dependence of the heat capacity obtained from the analysis of polypeptides energy fluctuations; the *solid line* corresponds to the derivative of the interpolating function of the transition energy on temperature. The figure is adopted from [27]

dependence, whereas the method based on differentiating of the energy of the system requires at least twice as many data points. Therefore the energy fluctuations method is more convenient for calculation of the heat capacity of polypeptides.

As it is seen from Fig. 5.14 the heat capacity of the Val$_{30}$ polypeptide acquires two peaks. Each peak is a result of certain structural transformation. The peak at higher temperature is due to the π-helix $\leftrightarrow$ random coil transition of the polypeptide. This is accompanied by the breaking of hydrogen bonds in the backbone of the polypeptide chain. The smaller peak at lower temperature can be explained by the dynamics of side chain radicals. At low temperature the side chain radicals of the Val$_{30}$ polypeptide form the ordered state in which they are aligned along the backbone of the polypeptide. With the increase of temperature the ordering of side chains becomes broken and the radicals rotate. The transition from the ordered state of side chains to a disordered state can be interpreted as a phase transition.

In order to clarify the nature of the structural transition involving the side chain radicals the structure parameter of the system is introduced as follows:

$$\vartheta_i = \langle \cos(\zeta(i) - \zeta_0(i)) \rangle. \tag{5.16}$$

Here $\zeta(i)$ is the angle between the radical of central amino acid (15th) and the radical of ith amino acid of Val$_{30}$ polypeptide. The definition of angle ζ is shown in the inset to Fig. 5.15. The angle $\zeta_0(i)$ is the value of $\zeta(i)$ for the equilibrated structure. In Fig. 5.15 is shown the dependence of the average value of the structure parameter ϑ_i calculated for different amino acids along the chain of Val$_{30}$ polypeptide at different temperatures. The structure parameter ϑ_i characterizes the relative alignment of the side chain radicals. For each i, ϑ_i has a limiting value at low and at high temperatures (two attractors). For temperatures 0–315 K, $\vartheta_i \approx 1$, while for

Fig. 5.15 Dependence of the average value of the structure parameter $\vartheta_i = \langle \cos(\zeta(i) - \zeta_0(i)) \rangle$ calculated for different amino acids along the chain of the Val_{30} polypeptide at different temperatures. Temperatures are given in Kelvin near each curve. Figure in the *inset* gives the definition of angle ζ, with ζ_0 being the value of ζ for equilibrated structure. The figure is adopted from [27]

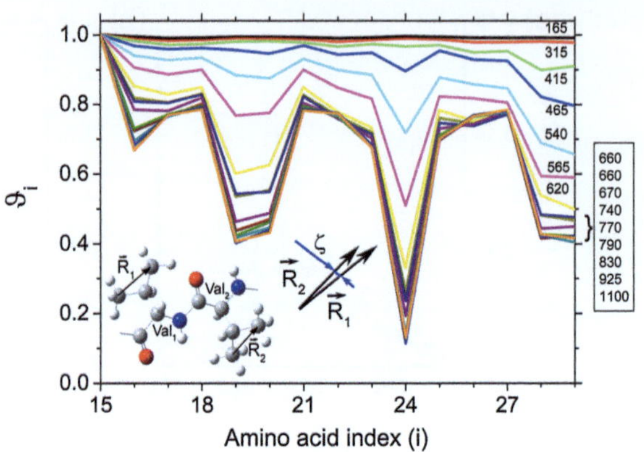

temperatures 660–1100 K, ϑ is approaching a certain limiting value (see Fig. 5.15. Note that the distribution of these limiting values oscillate as a function of amino acid index. These oscillations are due to the secondary structure motif of the polypeptide.

Temperatures at which ϑ_i are in between the two limiting values correspond to the temperature range of the structural transition. Thus, for the Val_{30} polypeptide this temperature range is equal to 465–565 K, which is the same range of temperature as it follows from the heat capacity analysis presented in Fig. 5.14.

Note that there is no second peak in the heat capacity dependence of Ala_{30} and Leu_{30} polypeptides (see Sects. 5.4.2 and 5.4.4). This can be explained as follows. In alanine polypeptides the side chain radicals are small and thus weakly bound. In leucine polypeptides the side chain radicals are larger and therefore the structural transition is shifted towards higher energies and takes place simultaneously with the π-helix $\leftrightarrow$ random coil of the backbone of the chain.

Let us now analyze dependence of the numerical error of the heat capacity on MD simulation time. For this purpose the following quantity is introduced:

$$\chi(\tau) = \sum_i \left(C_i(\tau) - C_i^{\text{ref}} \right)^2. \tag{5.17}$$

Here the summation is performed over all data points. $C_i(\tau)$ is the value of heat capacity obtained from MD simulation of duration τ. C_i^{ref} is the reference value of the heat capacity. In the present work it corresponds to the longest simulation (see Sect. 5.4.5). Assuming that $\chi(\tau)$ obeys the power law, (5.17) can be parameterized as follows.

$$\lg(\chi) = \alpha + \beta \lg(\tau), \tag{5.18}$$

where α and β are coefficients, τ is the simulation time. In Fig. 5.16 are shown the fluctuations of the heat capacity as a function of the simulation time (note the double decimal logarithm scale). From this figure it is seen that in the central part

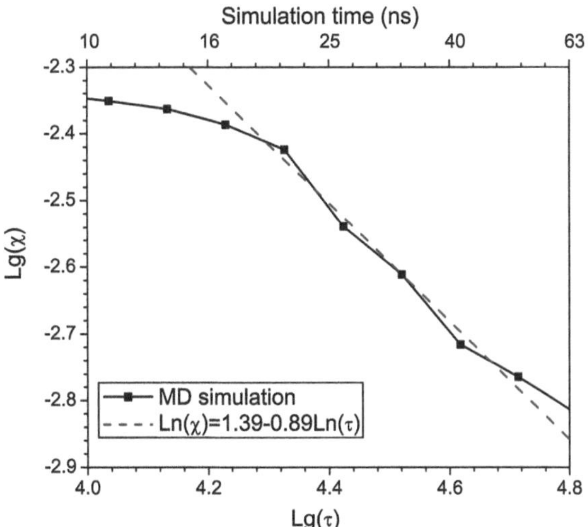

Fig. 5.16 Dependence of squared deviations of heat capacity χ, (5.17), as a function of simulation time τ. *Squares* show the results of MD simulations, while the *dashed line* corresponds to the linear fit, (5.18), of the relevant data. The figure is adopted from [27]

of the plot the dependence of $\lg(\chi)$ on $\lg(\tau)$ is linear and described by (5.18). The corresponding coefficients are: $\alpha = 1.39 \pm 0.29$, $\beta = -0.89 \pm 0.06$, leading to a conclusion that $\chi \sim 1/\tau$. The deviations from the linear behavior can be attributed to the following facts: at $\lg \tau < 4.2$ the simulation time is too short and insufficient for statistical description of the system; at $\lg \tau > 4.6$ the deviations arise due to the remaining statistical errors in the reference heat capacity curve C^{ref}.

5.4.4 π-Helix $\leftrightarrow$ Random Coil Transition in Leucine Polypeptide

Similar analysis was performed for leucine polypeptide consisting of 30 amino acids. Figure 5.17a shows the dependence of the transition energy on temperature which also shows a step-like dependence, characteristic for the first order like phase transition. In the case of leucine polypeptide in vacuo this transition corresponds to the π-helix $\leftrightarrow$ random coil transition.

In Fig. 5.17b is shown the heat capacity on temperature dependence calculated from the energy fluctuations (dots) and from the differentiation of the energy interpolating function (solid line). The use of energy interpolating function allows one to reduce the simulation time needed for the description of the phase transition. The heat capacity on temperature dependence obtained with this method can be used to identify the "temperature regions of interest". In these regions additional more systematic analysis of heat capacity should be performed with the use of energy fluctuations method.

Figure 5.17 shows that the phase transition in the leucine polypeptide is more pronounced than in the alanine and the valine polypeptides. For leucine the height of the peak C_0 is the largest, while the temperature range of the phase transition ΔT

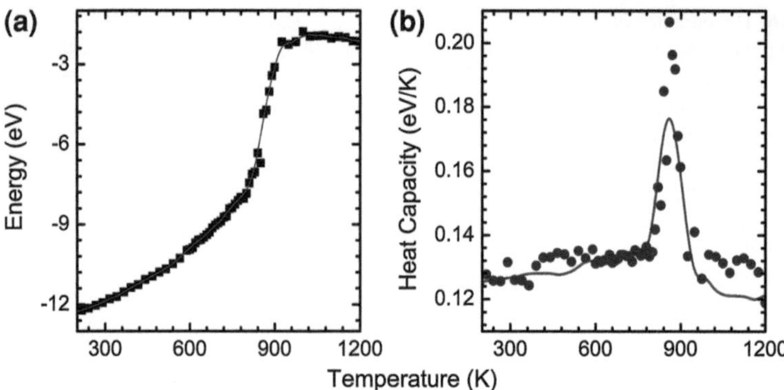

Fig. 5.17 Dependence of transition energy, (5.15), on temperature calculated for the Leu$_{30}$ polypeptide (**a**) and the corresponding dependence of the heat capacity on temperature (**b**). In **a**: the *squares* show the results of MD simulation, and the *solid line* shows the corresponding interpolating function. In **b**: the *dots* show the dependence of the heat capacity calculated from the energy fluctuations, the *solid line* corresponds to the derivative of the energy interpolating function with respect to temperature. The figure is adopted from [27]

are similar for the three polypeptides. Indeed, in both the alanine and the leucine polypeptide $\Delta T \simeq 80$ K, while $C_0 = 0.048$ eV/K and $C_0 = 0.078$ eV/K for alanine and leucine polypeptides respectively. This happens because leucine side chain radical is larger than in alanine and therefore the peak in the heat capacity is stronger expressed.

5.4.5 *Appendix: Parameters of MD Simulation*

In this appendix are presented the details of MD simulations performed in the present section. In the case of the Ala$_{30}$ polypeptide were performed 75 simulations within the temperature range between 15 and 1250 K, where simulations were performed with the temperature step size of 10 K for the 550–950 K and 25 K otherwise. The simulations in the temperature region 550–950 K were performed for 500 ns, and for 100 ns at other temperatures.

For the Val$_{30}$ polypeptide were performed 70 simulations within the temperature range between 15 and 1250 K. These simulations were performed with the temperature step size of 10 K for the 600–900 K and 25 K otherwise. Simulations in the temperature region 600–900 K were performed for 500 ns, and for 100 ns at other temperatures.

79 MD simulations of the Leu$_{30}$ polypeptide were performed, where 32 simulations were done for the temperature range of 590–910 K and 47 for temperatures 15–565 and 925–1500 K. The simulation time was 100 ns for the temperatures in the range 600–900 K and 20 ns otherwise.

To analyze the results of MD simulations cubic B-splines for energy interpolation were used, splines were smoothed over the whole temperature range as described in Ref. [39].

References

1. Zimm, B., & Bragg, J. (1959). Theory of the phase transition between helix and random coil in polypeptide chains. *Journal of Chemical Physics, 31*, 526–535.
2. Gibbs J., & DiMarzio E. (1959). Statistical mechanics of helix–coil transitions in biological macromolecules. *Journal of Chemical Physics, 30*, 271–282.
3. Lifson, S., & Roig, A. (1961). On the theory of helix–coil transition in polypeptides. *Journal of Chemical Physics, 34*, 1963–1974.
4. Schellman, J. A. (1958). The factors affecting the stability of hydrogen-bonded polypeptide structures in solution. *Journal of Chemical Physics, 62*, 1485–1494.
5. Lifson, S. (1964). Partition function of linear-chain molecules. *Journal of Chemical Physics, 40*, 3705–3710.
6. Poland, D., & Scheraga, H. A. (1966). Phase transitions in one dimension and the helix–coil transition in polyamino acids. *Journal of Chemical Physics, 45*, 1456–1463.
7. Ooi, T., & Oobatake, M. (1991). Prediction of the thermodynamics of protein unfolding: The helix–coil transition of poly(l-alanine). *Proceedings of the National Academy of Sciences of the United States of America, 88*, 2859–2863.
8. Gomez, J., Hilser, V. J., Xie, D., & Freire, E. (1995). The heat capacity of proteins. *Proteins: Structure, Function, and Genetics, 22*, 404–412.
9. Tobias, D. J., & Brooks, C. L., III (1991). Thermodynamics and mechanism of α helix initiation in alanine and valine peptides. *Biochemistry, 30*, 6059–6070.
10. Garcia, A. E., & Sanbonmatsu, K. Y. (2002). α-helical stabilization by side chain shielding of backbone hydrogen bonds. *Proceedings of the National Academy of Sciences of the United States of America, 99*, 2781–2787.
11. Nymeyer, H., & Garcia, A. E. (2003). Simulation of the folding equilibrium of α-helical peptides: A comparison of the generalized born approximation with explicit solvent. *Proceedings of the National Academy of Sciences of the United States of America, 100*, 13934–13939.
12. Irbäck, A., & Sjunnesson, F. (2004). Folding thermodynamics of three β-sheet peptides: A model study. *Proteins: Structure, Function, and Genetics, 56*, 110–116.
13. Shental-Bechor, D., Kirca, S., Ben-Tal, N., & Haliloglu, T. (2005). Monte Carlo studies of folding, dynamics and stability in α-helices. *Biophysical Journal, 88*, 2391–2402.
14. Kromhout, R. A., & Linder, B. (2001). Predictions of conformational enthalpy and heat capacity from the Zimm–Bragg theory. *The Journal of Physical Chemistry B, 105*, 4987–4991.
15. Chakrabartty, A., Kortemme, T., & Baldwin, R. L. (1994). Helix propensities of the amino acids measured in alanine-based peptides without helix-stabilizing side-chain interactions. *Protein Science, 3*, 843–852.
16. Go, M., Go, N., & Scheraga, H. A. (1970). Molecular theory of the helix–coil transition in polyamino acids. II. Numerical evaluation of s and σ for polyglycine and poly-l-alanine in the absence (for s and σ) and presence (for σ) of solvent. *Journal of Chemical Physics, 52*, 2060–2079.
17. Scheraga, H. A., Villa, J. A., & Ripoll, D. R. (2002). Helix–coil transitions re-visited. *Biophysical Chemistry, 101–102*, 255–265.
18. Scholtz, J. M., Marqusee, S., Baldwin, R. L., York, E. J., Stewart, J. M., Santoro, M., & Bolen, D. W. (1991). Calorimetric determination of the enthalpy change for the α- helix to coil transition of an alanine peptide in water. *Proceedings of the National Academy of Sciences of the United States of America, 88*, 2854–2858.

19. Lednev, I. K., Karnoup, A. S., Sparrow, M. C., & Asher, S. A. (2001). Transient UV raman spectroscopy finds no crossing barrier between the peptide α-helix and fully random coil conformation. *Journal of the American Chemical Society, 123*, 2388–2392.

20. Thompson, P. A., Eaton, W. A., & Hofrichter, J. (1997). Laser temperature jump study of the helix–coil kinetics of an alanine peptide interpreted with a 'kinetic zipper' model. *Biochemistry, 36*, 9200–9210.

21. Williams, S., Thimothy, R. G., Causgrove, P., Fang, K. S., Callender, R. H., Woodruff, W. H., & Dyer, R. B. (1996). Fast events in protein folding: Helix melting and formation in a small peptide. *Biochemistry, 35*, 691–697.

22. Finkelstein, A., & Ptitsyn, O. (2002). Protein physics. A course of lectures. Oxford: Elsevier Books.

23. Prabhu, N. V., & Sharp, K. A. (2005). Heat capacity in proteins. *Annual Review of Physical Chemistry, 56*, 521–548.

24. Shakhnovich, E. (2006). Protein folding thermodynamics and dynamics: Where physics, chemistry, and biology meet. *Chemical Reviews, 106*, 1559–1588.

25. Shea, J.-E., & Brooks, C. L. (2001). From folding theories to folding proteins: A review and assessment of simulation studies of protein folding and unfolding. *Annual Review of Physical Chemistry, 52*, 499–535.

26. Solov'yov, I., Yakubovich, A., Solov'yov, A., & Greiner, W. (2008). α-helix $\leftrightarrow$ random coil phase transition: Analysis of ab initio theory predictions. *European Physical Journal D, 46*, 227–240.

27. Yakubovich, A., Solov'yov, I., Solov'yov, A., & Greiner, W. (2008). Phase transitions in polypeptides: Analysis of energy fluctuations. *European Physical Journal D, 51*, 25–31.

28. Henriques, E., & Solov'yov, A. (2007). A rational method for probing macromolecules dissociation: The antibody-hapten system. *European Physical Journal D, 46*, 471–481.

29. Phillips, J. C., Braun, R., Wang, W., et al. (2005). Scalable molecular dynamics with NAMD. *Journal of Computational Chemistry, 26*, 1781–1802.

30. Freddolino, P., Arkhipov, A., Larson, S., McPherson, A., & Schulten, K. (2006). Molecular dynamics simulations of the complete satellite tobacco mosaic virus. *Structure, 14*, 437–449.

31. Rapaport, D. (2004). The art of molecular dynamics simulation. Cambridge: Cambridge University Press.

32. Frenkel, D., & Smit, B. J. (2002). Understanding molecular simulation: From algorithms to applications. Academic Press, San Diego.

33. Coffey, W., Kalmykov, Y., & Waldron, J. (2004). The Langevin equation, volume 14 of World Scientific in contemporary chemical physics. Singapore: World Scientific Publishing Co.

34. Reif, F. (1965). Fundamentals of statistical and thermal physics. New York: McGraw Hill.

35. MacKerell, A., Bashford, D., Bellott, M., Dunbrack, R., Evanseck, J., Field, M., Fischer, S., Gao, J., Guo, H., Ha, S., Joseph-McCarthy, D., Kuchnir, L., Kuczera, K., Lau, F., Mattos, C. M., Michnick, S., Ngo, T., Nguyen, D., Prodhom, B., Reiher, W., Roux, B., Schlenkrich, M., Smith, J., Stote, R., Straub, J., Watanabe, M., Wiorkiewicz-Kuczera, J., Yin, D., & Karplus, M. (1998). All-atom empirical potential for molecular modeling and dynamics studies of proteins. *The Journal of Physical Chemistry B, 102*, 3586–3616.

36. Sotomayor, M., Corey, D. P., & Schulten, K.(2005). In search of the hair-cell gating spring: Elastic properties of ankyrin and cadherin repeats. *Science, 13*, 669–682.

37. Gullingsrud, J., & Schulten, K. (2004). Lipid bilayer pressure profiles and mechanosensitive channel gating. *Biophysical Journal, 86*, 3496–3509.

38. Humphrey, W., Dalke, A., & Schulten, K. (1996). VMD—visual molecular dynamics. *Journal of Molecular Graphics, 14*, 33–38.

39. Press, W. H., et al. (2002). Numerical recipes in C++. Cambridge: Cambridge University Press.

40. Becke, A. (1988). Density-functional exchange-energy approximation with correct asymptotic-behavior. *Physical Review A, 38*, 3098–3100.

41. Lee, C., Yang, W., & Parr, R. (1988). Development of the colle-salvetti correlation-energy formula into a functional of the electron-density. *Physical Review B, 37*, 785–789.

42. Rubin, A. (2004). Biophysics: Theoretical biophysics. Moscow: Moscow University Press "Nauka".
43. Landau, L., & Lifshitz, E. (1959). Statistical physics. London: Pergamon Press.
44. Irbäck, A., Samuelsson, B., Sjunnesson, F., & Wallin, S. (2003). Thermodynamics of α- and β-structure formation in proteins. *Biophysical Journal, 85*, 1466–1473.
45. Nowak, C., Rostiashvilli, V. G., & Viglis, T. A. (1967). Globular structures of a helix–coil copolymer: Self-consistent treatment. *Journal of Chemical Physics, 46*, 4410–4426.
46. Ooi, T., Scott, R. A., Vanderkooi, G., & Scheraga, H. A. (1967). Conformational analysis of macromolecules. IV. Helical structures of poly-l-alanine, poly-l-valine, poly-β-methyl-l-aspartate, poly-γ-methyl-l-glutamate and poly-l-tyrosine. *Journal of Chemical Physics, 46*, 4410–4426.
47. Takano, M., Takahashi, T., & Nagayama, K. (1958). Helix–coil transition and 1/f fluctuation in a polypeptide. *Physical Review Letters, 80*, 5691–5694.
48. Wei, Y., Nadler, W., & Hansmann, U.H. (2006). Side-chain and backbone ordering in a polypeptide. *Journal of Chemical Physics, 125*, 164902–(1–5).
49. Alves, N. A., & Hansmann, U. H. (2002). Helix formation and folding in an artificial peptide. *Journal of Chemical Physics, 117*, 2337–2343.
50. Chen, Y., Zhou, Y., & Ding, J. (2007). The helix–coil transition revisited. *Proteins, 69*, 58–68.
51. Kemp, J. P., & Chen, Z. Y. (1998). Formation of helical states in wormlike polymer chains. *Physical Review Letters, 81*, 3880–3883.
52. Daggett, V., & Levitt, M. (1992). Molecular dynamics simulations of helix denaturation. *Journal of Molecular Biology, 223*, 1121–1138.

Chapter 6
Folding of Proteins in Aqueous Environment

6.1 Introduction

Proteins are biological polymers consisting of elementary structural units, amino acids. Being synthesized at ribosome, proteins are exposed to the cell interior, where they fold into their unique three dimensional structure. The correct folding of protein is of crucial importance for the protein's proper functioning. The current state-of-the-art in experimental and theoretical studies of the protein folding process are described in recent reviews and references therein [1–5].

In this chapter, a novel theoretical method for the description of the protein folding process is developed. The presented statistical mechanics model treats the folding $\leftrightarrow$ unfolding phase transition in single-domain proteins as a first-order phase transition in a finite system. The suggested method is based on the theory developed for the helix $\leftrightarrow$ coil transition in polypeptides discussed in the previous Chaps. 3–5 of the thesis. A way to construct a parameter-free partition function for a system experiencing α-helix $\leftrightarrow$ random coil phase transition in vacuo was studied in Chap. 4. In Chap. 5 were calculated the potential energy surfaces (PES) of polyalanines of different lengths with respect to their twisting degrees of freedom. This was done within the framework of classical molecular mechanics. The calculated PES were then used to construct a parameter-free partition function of a polypeptide and to derive various thermodynamical characteristics of alanine polypeptides as a function of temperature and polypeptide length.

In this chapter the partition function of a protein in vacuo is constructed, which is the further generalization of the formalism developed in Sect. 4.4 of the thesis, accounting for folded, unfolded and prefolded states of the protein. This way of the construction of the partition function is consistent with nucleation–condensation scenario of protein folding, which is a very common scenario for globular proteins [6] and implies that at the early stage of protein folding the native-like hydrophobic nucleus of protein is formed, while at the later stages of the protein folding process all rest of amino acids also attain the native-like conformation. This chapter is based on the work published in [7, 8].

A. V. Yakubovich, *Theory of Phase Transitions in Polypeptides and Proteins*,
Springer Theses, DOI: 10.1007/978-3-642-22592-5_6,
© Springer-Verlag Berlin Heidelberg 2011

For the correct description of the protein folding in water environment it is of primary importance to consider the interactions between the protein and the solvent molecules. The hydrophobic interactions are known to be the most important driving forces of protein folding [9]. In the thesis a way of how one can construct the partition function of the protein accounting for the interactions with solvent, i.e., accounting for the hydrophobic effect is presented. The most prominent feature of the developed approach is that it is developed for concrete systems in contrast to various generalized and toy-models of protein folding process.

The hydrophobic interactions in the system are treated using the statistical mechanics formalism developed in [10] for the description of the thermodynamical properties of the solvation process of aliphatic and aromatic hydrocarbons in water. However, accounting solely for hydrophobic interactions is not sufficient for the proper description of the energetics of all conformational states of the protein and one has to take electrostatic interactions into account. In the present work the electrostatic interactions are treated within a similar framework as described in [11].

The developed statistical mechanics model of protein folding was applied to two globular proteins, namely, staphylococcal nuclease and metmyoglobin. These proteins have simple two-stage-like folding kinetics and demonstrate two folding $\leftrightarrow$ unfolding transitions, refereed as heat and cold denaturation [12, 13]. The comparison of the results of the theoretical model with that of the experimental measurements shows the applicability of the suggested formalism for an accurate description of various thermodynamical characteristics in the system, e.g., heat denaturation, cold denaturation, increase of the reminiscent heat capacity of the unfolded protein, etc.

This chapter is organized as follows. In Sect. 6.2.1 the formalism for the construction of the partition function of the protein in water environment is presented and the assumptions made on the system's properties are justified. In Sect. 6.3 the results obtained with the theoretical model for the description of folding $\leftrightarrow$ unfolding transition in staphylococcal nuclease and metmyoglobin are discussed.

6.2 Theoretical Methods

6.2.1 Partition Function of a Protein

To study thermodynamic properties of the system one needs to investigate its potential energy surface with respect to all the degrees of freedom. For the description of macromolecular systems, such as proteins, efficient model approaches are necessary.

The most relevant degrees of freedom in the protein folding process are the twisting degrees of freedom along its backbone chain as discussed in Chaps. 3, 4. These degrees of freedom are defined for each amino acid of the protein except for the boundary ones and are described by two dihedral angles φ_i and ψ_i (for definition of φ_i and ψ_i see Fig. 3.1).

A Hamiltonian of a protein is constructed as a sum of the potential, kinetic and vibrational energy terms. Assuming the harmonic approximation for the stiff degrees of freedom it is possible to derive in analogy to (4.17) the following expression for the partition function of a protein in vacuo being in a particular conformational state j:

$$Z_j = A_j (kT)^{3N-3-\frac{l_s}{2}} \int_{\varphi \in \Gamma_j} \cdots \int_{\psi \in \Gamma_j} e^{-\epsilon_j(\{\varphi,\psi\})/kT}\, d\varphi_1 \cdots d\varphi_n d\psi_1 \cdots d\psi_n, \quad (6.1)$$

where T is the temperature, k is the Boltzmann constant, N is the total number of atoms in the protein, l_s is the number of soft degrees of freedom, A_j is defined as follows:

$$A_j = \left[\frac{V_j \cdot M^{3/2} \cdot \sqrt{I_j^{(1)} I_j^{(2)} I_j^{(3)}}\, \prod_{i=1}^{l_s} \sqrt{\mu_i^{s(j)}}}{(2\pi)^{l_s} 2\pi \hbar^{3N} \prod_{i=1}^{3N-6-l_s} \omega_i^{(j)}} \right]. \quad (6.2)$$

A_j is a factor which depends on the mass of the protein M, its three main momenta of inertia $I_j^{(1,2,3)}$, specific volume V_j, the frequencies of the stiff normal vibrational modes $\omega_i^{(j)}$ and on the generalized masses $\mu_i^{s(j)}$ corresponding to the soft degrees of freedom (see Sect. 4.3). ϵ_i in (6.1) describes the potential energy of the system corresponding to the variation of soft degrees of freedom. Integration in (6.1) is performed over a certain part of a phase space of the system (a subspace Γ_j) corresponding to the soft degrees of freedom φ and ψ. The form of the partition function in (6.1) allows one to avoid the multidimensional integration over the whole coordinate space and to reduce the integration only to the relevant parts of the phase space. ϵ_j in (6.1) denotes the potential energy surface of the protein as a function of twisting degrees of freedom in the vicinity of protein's conformational state j. Note that in general the proper choice of all the relevant conformations of protein and the corresponding set of Γ_j is not a trivial task.

One can expect that the factors A_j in (6.1) depend on the chosen conformation of the protein. However, due to the fact that the values of specific volumes, momenta of inertia and frequencies of normal vibration modes of the system in different conformations are expected to be close [14], the values of A_j in all conformations become nearly equal, at least in the zero order harmonic approximation, i.e. $A_j \equiv A$. Another simplification of the integration in (6.1) comes from the statistical independence of amino acids: within each conformational state j all amino acids can be treated statistically independently, i.e. the particular conformational state of ith amino acid characterized by angles $\varphi_i \in \Gamma_j$ and $\psi_i \in \Gamma_j$ does not influence the potential energy surface of all other amino acids, and vice versa. This assumption is well applicable for rigid conformational states of the protein such as native state. For the native state of a protein all atoms of the molecule move in harmonic potential in the vicinity of their equilibrium positions. However, in unfolded states of the protein the flexibility of the backbone chain leads to significant variations of the distances between atoms, and consequently to a significant variation of interactions between atoms. Accurate accounting (both analytical and computational) for the interactions between distant atoms in the unfolded state of a protein is extremely difficult (see Ref. [15] for analytical treatment of interactions in unfolded states of a protein). In this work all amino acids in unfolded state of a protein are considered as moving in the identical mean field created by all the amino acids. The corrections to this approximation is left for further considerations.

With the above mentioned assumptions the partition function of a protein Z_p (without any solvent) reads as:

$$Z_p = A \cdot (kT)^{3N-3-\frac{l_s}{2}} \sum_{j=1}^{\xi} \prod_{i=1}^{a} \int_{-\pi}^{\pi} \int_{-\pi}^{\pi} \exp\left(-\frac{\epsilon_i^{(j)}(\varphi_i, \psi_i)}{kT} \right) d\varphi_i \, d\psi_i, \qquad (6.3)$$

where the summation over j includes all ξ statistically relevant conformations of the protein, a is the number of amino acids in the protein and $\epsilon_i^{(j)}$ is the potential energy surface as a function of twisting degrees of freedom φ_i and ψ_i of the ith amino acid in the jth conformational state of the protein. The exact construction of $\epsilon_a^{(j)}(\varphi_i, \psi_i)$ for various conformational states of a particular protein will be discussed below. The angles φ and ψ are considered as the only two soft degrees of freedom in each amino acid of the protein, and therefore the total number of soft degrees of freedom of the protein $l_s = 2a$.

Partition function in (6.3) can be further simplified if one assumes (i) that each amino acid in the protein can exist only in two conformations: the native state conformation and the random coil conformation; (ii) the potential energy surfaces for all the amino acids are identical. This assumption is applicable for both the native and the random coil state. It is not very accurate for the description of thermodynamical properties of single amino acids, but is reasonable for the treatment of thermodynamical properties of the entire protein. The judgment of the quality of this assumption could be made on the basis of comparison of the results obtained with its use with experimental data. Such comparison is performed in Sect. 6.3 of the thesis.

Amino acids in a protein being in its native state vibrate in a steep harmonic potential. vibrate in a steep harmonic potential. The potential energy profile of an amino acid in the native conformation should not be very sensitive to the type of amino acid and thus can be taken as, e.g., the potential energy surface for an alanine amino acid in the α-helix conformation (see Fig. 5.2). Using the same arguments the potential energy profile for an amino acid in unfolded protein state can be approximated by e.g. the potential of alanine in the unfolded state of alanine polypeptide (see Sect. 5.3.2 for discussion of alanine's potential energy surfaces). Indeed, for an unfolded state of a protein it is reasonable to expect that once neglecting the long-range interactions all the differences in the potential energy surfaces of various amino acids arise from the steric overlap of the amino acids's radicals. This is clearly seen on alanine's potential energy surface at values of $\varphi > 0°$ presented in Fig. 5.2. But the part of the potential energy surface at $\varphi > 0°$ gives a minor contribution to the entropy of amino acid at room temperature. This fact allows one to neglect all the differences in potential energy surfaces for different amino acids in an unfolded protein, at least in the zero order approximation. This assumption should be especially justified for proteins with the rigid helix-rich native structure. The staphylococcal nuclease, which is studied here has definitely high α-helix content. Another argument which allows to justify the assumption made for a wider family of proteins is the rigidity of the protein's native structure. Below, the assumptions made are validated by the comparison of the results of the theoretical model with the experimental data for α/β rich protein metmyoglobin obtained in [13].

For the description of the folding $\leftrightarrow$ unfolding transition in small globular proteins obeying simple two-state-like folding kinetics the protein is considered to exist in one of three states: completely folded state, completely unfolded state and partially folded state where some amino acids from the flexible regions with no prominent secondary structure are in the unfolded state, while other amino acids are in the folded conformation. With this assumption the partition function of the protein reads as:

$$Z_p = Z_0 + \sum_{i=a-\kappa}^{a} \frac{\kappa!}{(i-(a-\kappa))!(a-i)!} Z_i, \tag{6.4}$$

where Z_i is defined in (6.1), Z_0 is the partition function of the protein in completely unfolded state, a is the total number of amino acids in a protein and κ is the number of amino acids in flexible regions. The factorial term in (6.4) accounts for the states in which various amino acids from flexible regions independently attain the native conformation. The summation in (6.4) is performed over all partially folded states of the protein, where $a - \kappa$ is the minimal possible number of amino acids being in the folded state. The factorial term describes the number of ways to select $i - (a - \kappa)$ amino acids from the flexible region of the protein consisting of κ amino acids attaining native-like conformation.

Finally, the partition function of the protein in vacuo has the following form:

$$Z_p = \tilde{Z}_p \cdot A(kT)^{3N-3-a}, \tag{6.5}$$

where

$$\tilde{Z}_p = Z_u^a + \sum_{i=a-\kappa}^{a} \frac{\kappa! Z_b^i Z_u^{a-i} \exp\left(i \cdot E_0/kT\right)}{(i-(a-\kappa))!(a-i)!} \tag{6.6}$$

$$Z_b = \int_{-\pi}^{\pi}\int_{-\pi}^{\pi} \exp\left(-\frac{\epsilon_b(\varphi,\psi)}{kT}\right) d\varphi\, d\psi \tag{6.7}$$

$$Z_u = \int_{-\pi}^{\pi}\int_{-\pi}^{\pi} \exp\left(-\frac{\epsilon_u(\varphi,\psi)}{kT}\right) d\varphi\, d\psi. \tag{6.8}$$

Here was omitted the trivial factor describing the motion of the protein center of mass, which is of no significance for the problem considered, $\epsilon_b(\varphi, \psi)$ (b stands for *bound*) is the potential energy surface of an amino acid in the native conformation and $\epsilon_u(\varphi, \psi)$ (u stands for *unbound*) is the potential energy surface of an amino acid in the random coil conformation. The potential energy profile of an amino acid is calculated as a function of its twisting degrees of freedom φ and ψ. ϵ_b^0 and ϵ_u^0 denote the global minima on the potential energy surfaces of an amino acid in folded and in unfolded conformations respectively. The potential energy of an amino acid then reads as $\epsilon_{u,b}^0 + \epsilon_{u,b}(\varphi, \psi)$. E_0 in (6.6) is defined as the energy difference between the global energy minima of the amino acid potential energy surfaces corresponding to the folded and unfolded conformations, i.e. $E_0 = \epsilon_u^0 - \epsilon_b^0$. The potential energy surfaces for amino acids as functions of angles φ and ψ were calculated and thoroughly analyzed in Chap. 5.

In nature proteins perform their function in the aqueous environment. Therefore the correct theoretical description of the folding $\leftrightarrow$ unfolding transition in water environment should account for solvent effects.

6.2.2 Partition Function of a Protein in Water Environment

In this section E_0 is evaluated and the partition function for the protein in water environment is constructed.

The partition function of the infinitely diluted solution of proteins Z can be constructed as follows:

$$Z = \sum_{j=1}^{\xi} \tilde{Z}_p^{(j)} Z_W^{(j)}, \tag{6.9}$$

where $Z_W^{(j)}$ is the partition function of all water molecules in the jth conformational state of a protein and $\tilde{Z}_p^{(j)}$ is the partition function of the protein in its jth conformational state, in which the factor describing the contribution of stiff degrees of freedom in the system is further omitted. This is done in order to simplify the expressions, because stiff degrees of freedom provide a constant contribution to the heat capacity of the system since the heat capacity of the ensemble of harmonic oscillators is constant. Below for the simplicity of notations is put $\tilde{Z}_p \equiv Z_p$.

There are two types of water molecules in the system: (i) molecules in pure water and (ii) molecules interacting with the protein. Only the water molecules being in the vicinity of the protein's surface can be considered as involved in the folding $\leftrightarrow$ unfolding transition, because they are affected by the variation of the hydrophobic surface of a protein. This surface is equal to the protein's solvent accessible surface area (SASA) of the hydrophobic amino acids. The number of interacting molecules is proportional to SASA and include only the molecules from the first protein's solvation shell. This area depends on the conformation of the protein. The main contribution to the energy of the system caused by the variation of the protein's SASA is associated with the side-chain radicals of amino acids because the contribution to the free energy associated with solvation of protein's backbone is small [16]. Thus, main attention is payed to the accounting for the SASA change arising due to the solvation of side chain radicals.

All water molecules are treated as statistically independent, i.e. the energy spectra of the states of a given molecule and its vibrational frequencies do not depend on a particular state of all other water molecules. Thus, the partition function of the whole system Z can be factorized and reads as:

$$Z = \sum_{j=1}^{\xi} Z_p^{(j)} Z_s^{Y_c(j)} Z_w^{N_t - Y_c(j)}, \tag{6.10}$$

where ξ is the total number of states of a protein, Z_s is the partition function of a water molecule affected by the interaction with the protein and Z_w is the partition function of a water molecule in pure water. $Y_c(j)$ is the number of water molecules interacting with the protein in the jth conformational state. N_t is the total number of water molecules in the system. To simplify the expressions, the water molecules that do not interact with the protein in any of its conformational states are not accounted in further equations, i.e. $N_t = \max_j\{Y_c(j)\}$.

The construction of the partition function of water is based on the formalism developed in [10]. Here only to the most essential details of that work are presented. The partition function of a water molecule in pure water reads as:

$$Z_w = \sum_{l=0}^{4} \left[\xi_l f_l \exp(-E_l/kT)\right], \tag{6.11}$$

where the summation is performed over five possible states of a water molecule (the states in which water molecule has $4, 3, 2, 1$ or 0 hydrogen bonds with the neighboring molecules). E_l are the energies of these states and ξ_l are the combinatorial factors being equal to $1, 4, 6, 4, 1$ for $l = 0, 1, 2, 3, 4$, respectively. They describe the number of choices to form a given number of hydrogen bonds. f_l in (6.11) describes the contribution due to the partition function arising due to the translation and libration oscillations of the molecule. In the harmonic approximation f_l are equal to:

$$f_l = \left[1 - \exp(-h\nu_l^{(T)}/kT)\right]^{-3} \left[1 - \exp(-h\nu_l^{(L)}/kT)\right]^{-3}, \tag{6.12}$$

where $\nu_l^{(T)}$ and $\nu_l^{(L)}$ are translation and libration motions frequencies of a water molecule in its lth state, respectively. These frequencies are calculated in Ref. [10] and are given in Table 6.1. The contribution of the internal vibrations of water molecules is not included in (6.11) because the frequencies of these vibrations are practically not influenced by the interactions with surrounding water molecules.

Table 6.1 Parameters of the partition function of water according to [10]

Number of hydrogen bonds	0	1	2	3	4
Energy level, E_i (kcal/mol)	6.670	4.970	3.870	2.030	0
Energy level, E_i^s (kcal/mol)	6.431	4.731	3.631	1.791	−0.564
Translational frequencies, $v_i^{(T)}$ (cm^{-1})	26	86	61	57	210
Librational frequencies, $v_i^{(L)}$ (cm^{-1})	197	374	500	750	750

The partition function of a water molecule from the protein's first solvation shell reads as:

$$Z_s = \sum_{l=0}^{4} \left[\xi_l\, f_l \exp(-E_l^s / kT) \right], \tag{6.13}$$

where f_l are defined in (6.12) and E_l^s denotes the energy levels of a water molecule interacting with aliphatic hydrocarbons of protein's amino acids. Values of energies E_l^s are given it Table 6.1. For simplicity all side-chain radicals of a protein are treated as aliphatic hydrocarbons because most of the protein's hydrophobic amino acids consist of aliphatic-like hydrocarbons. It is possible to account for various types of side chain radicals by using the experimental results of the measurements of the solvation free energies of amino acid radicals from Ref. [17] and associated works. However, this correction will imply the reparametrization of the theory presented in [10] and will lead to the introduction of $\sim 20 \cdot 5$ additional parameters. Here such a task is not performed since this kind of improvement of the theory would smear out the understanding of the principal physical factors underlying the protein folding $\leftrightarrow$ unfolding transition.

The theoretical model also accounts for the electrostatic interaction of protein's charged groups with water. The presence of electrostatic field around the protein leads to the reorientation of H_2O molecules in the vicinity of charged groups due to the interaction of dipole moments of the molecules with the electrostatic field. The additional factor arising in the partition function (6.11) of the molecules reads as:

$$Z_E = \left(\frac{1}{4\pi} \int \exp\left(-\frac{E \cdot d \cos\theta}{kT} \right) \sin\theta\, d\theta\, d\varphi \right)^{\alpha}, \tag{6.14}$$

where E is the strength of the electrostatic field, d is the absolute value of the H_2O molecule dipole moment, α is the ratio of the number of water molecules that interact with the electrostatic field of the protein (N_E) to the number of water molecules interacting with the surface of the amino acids from the inner part of the protein while they are exposed to water when the protein is being unfolded (N_w), i.e. $\alpha = N_E / N_w$. Note that the effects of electrostatic interaction turn out to be more pronounced in the folded state of the protein. This happens because in the unfolded state of a protein opposite charges of amino acid's radicals are in average closer in space due to the flexibility of the backbone chain, while in the folded state the positions of the charges are fixed by the rigid structure of a protein.

Integrating (6.14) allows to write the factor Z_E for the partition function of a single H_2O molecule in pure water in the form:

$$Z_E = \left(\frac{kT \sinh\left[\frac{Ed}{kT}\right]}{Ed} \right)^{\alpha}.$$ (6.15)

This equation shows how the electrostatic field enters the partition function. In general, E depends on the position in space with respect to the protein. However, here this dependence is neglected and instead the parameter E is treated as an average, characteristic electrostatic field created by the protein.

Let us denote by N_s the number of water molecules interacting with the proteins surface in its folded state i.e. $N_t = N_s + N_w$; where N_t is defined in (6.10). The number of water molecules interacting with the protein (Y_c) can be considered as being linearly dependent on the number of amino acids being in the unfolded conformation, i.e. $Y_c = N_s + i N_w / a$, where i is the number of the amino acids in the unfolded conformation and a is the total number of amino acids in the protein. Thus, the partition function (6.10) with the accounting for the factor (6.15) reads as:

$$Z = Z_s^{N_s} \cdot \sum_{j=1}^{\xi} \left(Z_b Z_w^{N_w/a} Z_E^{N_w/a} \exp(i \cdot E_0/kT) \right)^{i(j)} \left(Z_u Z_s^{N_w/a} \right)^{a-i(j)},$$ (6.16)

where $i(j)$ denotes the number of the amino acids being in the folded conformation when the protein is in the jth conformational state. Accounting for the statistical factors for amino acids being in the folded and unfolded states, similarly to how it was done for the vacuum case (see (6.6)), one derives from (6.16) the following final expression:

$$Z = (Z_s)^{N_s}$$
$$\times \left[Z_u^a Z_s^{N_w} + \sum_{i=a-\kappa}^{a} \frac{\kappa! \exp(i \cdot E_0/kT)}{(i-(a-\kappa))!(a-i)!} \left(Z_b Z_w^{N_w/a} Z_E^{N_w/a} \right)^{i} (Z_u Z_s^{N_w/a})^{a-i} \right],$$ (6.17)

where the term in the square brackets accounts for all statistically significant conformational states of the protein.

Having constructed the partition function of the system one can evaluate with its use all thermodynamic characteristics of the system, such as e.g. entropy, free energy, heat capacity, etc. The free energy (F) and heat the capacity (c) of the system can be calculated from the partition function as follows:

$$F(T) = -kT \ln Z(T),$$ (6.18)

$$c(T) = -T \frac{\partial^2 F(T)}{\partial T^2}.$$ (6.19)

In this chapter the dependence of protein's heat capacity on temperature is analyzed and the predictions of the model with available experimental data are compared.

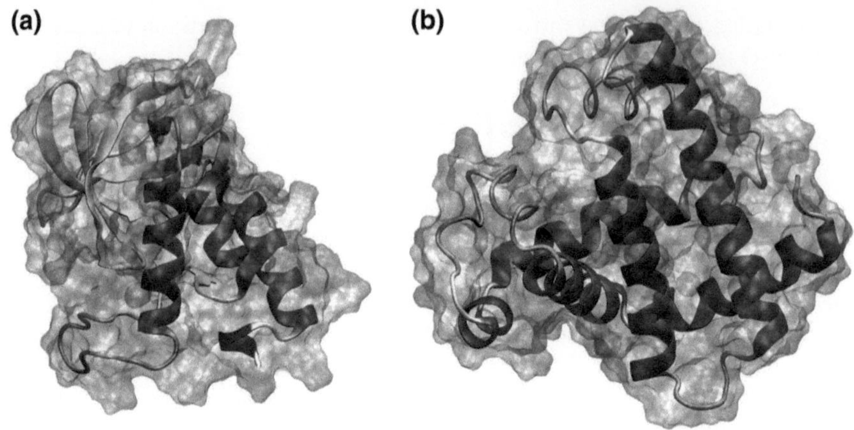

Fig. 6.1 **a** Structure of staphylococcal nuclease (PDB ID 1EYD [27]), and **b** horse heart metmyoglobin (PDB ID 1YMB [28]). Images have been rendered using VMD program [29]. The figure is adopted from [8]

6.3 Results and Discussion

In this section the dependencies of the heat capacity on temperature is calculated for two globular proteins metmyoglobin and staphylococcal nuclease and the results are compared with experimental data from [12, 13].

The structures of metmyoglobin and staphylococcal nuclease proteins are shown in Fig. 6.1. These are relatively small globular proteins consisting of $\sim$150 amino acids. Under certain experimental conditions (salt concentration and pH) the metmyoglobin and the staphylococcal nuclease experience two folding $\leftrightarrow$ unfolding transitions, which induce two peaks in the dependency of heat capacity on temperature (see further discussion). The peaks at lower temperature are due to the cold denaturation of the proteins. The peaks at higher temperatures arise due to the ordinary folding $\leftrightarrow$ unfolding transition. The availability of experimental data for the heat capacity profiles of the mentioned proteins, the presence of the cold denaturation and simple two-stage-like folding kinetics are the reasons for selecting these particular proteins as case studies for the verification of the developed theoretical model.

6.3.1 Heat Capacity of Staphylococcal Nuclease

Staphylococcal or micrococcal nuclease (S7 Nuclease) is a relatively nonspecific enzyme that digests single-stranded and double-stranded nucleic acids, but is more active on single-stranded substrates [18]. This protein consists of 149 amino acids. It's structure is shown in Fig. 6.1a.

To calculate the SASA of staphylococcal nuclease in the folded state the 3D structure of the protein was taken from the Protein Data Bank [19] (PDB ID 1EYD). Using CHARMM27 [20] forcefield and NAMD program [21] the structural optimization of the protein was performed and SASA was calculated with the solvent probe radius 1.4 Å.

The value of SASA of the side-chain radicals in the folded protein conformation is equal to $S_f = 6,858$ Å^2. In order to calculate SASA for an unfolded protein state, the value of all angles φ and ψ were put equal to $180°$, corresponding to a fully stretched conformation. Then, the optimization of the structure with the fixed angles φ and ψ was performed. The optimized geometry of the stretched molecule has a minor dependence on the value of dielectric susceptibility of the solvent, therefore the value of dielectric susceptibility was chosen to be equal to 20, in order to mimic the screening of charges by the solvent. SASA of the side-chain radicals in the stretched conformation of the protein is equal to $S_u = 15,813$ Å^2.

The change of the number of water molecules those interacting with the protein due to the unfolding process can be calculated as follows:

$$N_w = (S_u - S_f)n^{2/3}, \tag{6.20}$$

where $S_u = 15,813$ Å^2 and $S_f = 6,858$ Å^2 are the SASA of the protein in unfolded and in folded conformations, respectively and n is the density of the water molecules. The volume of one mole of water is equal to $18\,\text{cm}^3$, therefore $n \approx 30\,\text{Å}^{-3}$.

To account for the effects caused by the electrostatic interaction of water molecules with the charged groups of the protein it is necessary to evaluate the strength of the average electrostatic field E in (6.15). The strength of the average field can be estimated as $E \cdot d = kT$, where d is the dipole moment of a water molecule, k is Boltzmann constant and $T=300\,\text{K}$ is the room temperature. According to this estimate the energy of characteristic electrostatic interaction of water molecules is equal to the thermal energy per degree of freedom of a molecule.

The total number of water molecules that interact with the electrostatic field of the protein can be estimated from the known Debye screening length of a charge in electrolyte as follows:

$$N_E = N_q \frac{4\pi\rho}{3}\lambda_d^3, \tag{6.21}$$

where N_q is the number of charged groups in the protein, ρ is the density of water and λ is the Debye screening length. Debye screening length of the symmetric electrolyte can be calculated as follows [22]:

$$\lambda_d = \sqrt{\frac{\epsilon\epsilon_0 kT}{2N_A e^2 I}}, \tag{6.22}$$

where ϵ_0 is the permittivity of free space, ϵ is the dielectric constant, N_A is the Avogadro number, e is the elementary charge and I is the ionic strength of the electrolyte.

The experiments on denaturation of staphylococcal nuclease and metmyoglobin were performed in 100 mM ion buffer of sodium chloride and 10 mM buffer of sodium acetate respectively [12, 13]. The Debye screening length in water with 10 and 100 mM concentration of ions is $\lambda_d = 30$ Å and $\lambda_d = 10$ Å at room temperature respectively.

The described method allows to estimate the number of water molecules (N_E) interacting with electric filed created by the charged groups of a protein. It should be considered as qualitative estimate since the average electric field was assumed to be constant within a sphere of the radius λ_d, but in fact it experiences some variations. Thus, at the distances ~ 15 Åfrom the point charge the interaction energy of a H_2O molecule with the electric field becomes equal to ~ 0.02 kT (for this estimate the linear growing distance-dependent dielectric susceptibility $\epsilon = 6R$ was used as derived in [23] for the atoms fully exposed to the solvent). However, the more accurate analysis accounting for the spatial variation of the electric field will not change significantly the results of the analysis reported here, because it is based on the physically correct picture of the effect and the realistic values of all the physical quantities. At physiological conditions staphylococcal nuclease has eight charged residues [24]. The value of α for this protein varies within the interval from 1.29 to 31.27 for $\lambda_d \in [10..30]$ Å. In the numerical analysis the characteristic value of α equal to 2.5 was used.

Note that number of molecules interacting with the electrostatic field N_E and the strength of the electrostatic field E should be considered as the *effective* parameters of the model. In this work the accurate accounting for the spatial dependence of the electrostatic field is not performed. Instead, the parameters α and E are introduced. These parameters can be interpreted as effective values of the number of H_2O molecules and the strength of the electrostatic field correspondingly. Mention, that the number of water molecules α and the strength of the field E are not independent parameters of the model because by choosing the higher value of E and smaller value of α or vice versa one can derive the same heat capacity profile.

The dependencies of the heat capacity profiles on the values of the parameters α and E are not investigated in the present work. Below the investigation of the dependence of the protein heat capacity on the energy E_0 is performed at the fixed value of α and E equal to 2.5 and 0.58 kcal/mol respectively.

An important parameter of the model is the energy difference between the two states of the protein normalized per one amino acid, E_0 introduced in (6.6). This parameter describes both the energy loss due to the separation of the hydrophobic groups of the protein which attract in the native state of the protein due to Van-der-Waals interaction and the energy gain due to the formation of Van-der-Waals interactions of hydrophobic groups of the protein with H_2O molecules in the protein's unfolded state. Also, the difference of the electrostatic energy of the system in the folded and unfolded states is accounted for in E_0. The difference of the electrostatic energy may depend on various characteristics of the system, such as concentration of ions in the solvent and its pH, on the exact location of the charged sites in the native conformation of the protein and on the probability distribution of distances between charged amino acids in the unfolded state. Thus, exact calculation

Table 6.2 Values of E_0 for staphylococcal nuclease at different values of pH of the solvent

pH value	7.0	5.0	4.5	3.88	3.23
E_0 (kcal/mol)	0.789	0.795	0.803	0.819	0.890

of E_0 is rather difficult. It is a separate task which is not addressed in the present thesis. Instead, in the current study the energy difference between the two phases of the protein is considered as a parameter of the model. E_0 is treated as being dependent on external properties of the system, in particular on the pH value of the solution.

Another characteristic of the protein folding $\leftrightarrow$ unfolding transition is its cooperativity. In the model it is described by the parameter κ in (6.4). κ describes the number of amino acids in the flexible regions of the protein. The staphylococcal nuclease possesses a prominent two-stage folding kinetics, therefore only 5–10% of amino acids is in the protein's flexible regions. Thus, the value of κ for this protein is small. It can be estimated as being equal to $149 \cdot 7\% \approx 10$ amino acids.

The values of E_0 for staphylococcal nuclease at different values of pH are given in Table 6.2.

For the analysis of the variation of the thermodynamic properties of the system during the folding process one can omit all the contributions to the free energy of the system that do not alter significantly in the temperature range between $-50°$ C and $150°$. Therefore, from the expression for the total free energy of the system F one can subtract all slowly varying contributions F_0 as follows:

$$\delta F = F - F_0 = -(kT \ln Z - kT \ln Z_0) = -kT \ln \left(\frac{Z}{Z_0} \right). \tag{6.23}$$

From (6.23) follows that the subtraction of F_0 corresponds to the division of the total partition function Z by the partition function of the subsystem (Z_0) with slowly varying thermodynamical properties. Therefore, in order to simplify the expressions, one can divide the partition function in (6.17) by the partition function of fully unfolded conformation of a protein (by $Z_u^a Z_s^{N_w}$) and by the partition function of N_s free water molecules (by $Z_w^{N_s}$). Thus, (6.17) can be rewritten as follows:

$$Z = \left(\frac{Z_s}{Z_w} \right)^{N_s} \left(1 + \sum_{i=a-\kappa}^{a} \frac{\kappa! \exp \left(i \cdot E_0/kT \right)}{(i - (a - \kappa))!(a - i)!} \left(\frac{Z_b}{Z_u} \right)^i \left(\frac{Z_w Z_E}{Z_s} \right)^{i N_w/a} \right). \tag{6.24}$$

With the use of (6.19) on can calculate the heat capacity of the system as follows:

$$c(T) = A + B(T - T_0) - T \frac{\partial^2 F(T)}{\partial T^2}, \tag{6.25}$$

where the factors A and B are responsible for the absolute value and the inclination of the heat capacity curve respectively. These factors account for the contribution of stiff harmonic vibrational modes in the system (factor A) and for the unharmonic correction to these vibrations (factors B and T_0). The contribution of protein's stiff

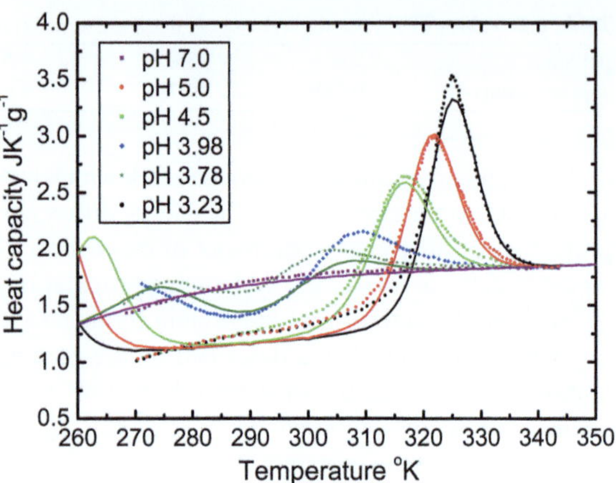

Fig. 6.2 Dependencies of the heat capacity on temperature for staphylococcal nuclease (see Fig. 6.1a) at different values of pH. *Solid lines* show results of the calculation, while *symbols* present experimental data from Ref. [12]. The figure is adopted from [8]

vibrational modes and the heat capacity of the fully unfolded conformation of protein is also included into these factors. In the numerical analysis the values of A, B and T_0 were adjusted in order to match experimental measurements. However, factors A, B and T_0 should not be considered as parameters of the model since their values are not related to the thermodynamic characteristics of the folding $\leftrightarrow$ unfolding transition and depend not entirely on the properties of the protein but also on the properties of the solution, protein and ion concentrations, etc.

In the calculations for staphylococcal nuclease the values of $A = 1.25\,\mathrm{J\,K^{-1}\,g^{-1}}$, $B = 6.25 \times 10^{-3}\,\mathrm{J\,K^{-2}\,g^{-1}}$ and $T_0 = 323°\,\mathrm{K}$ were used in (6.25).

The dependencies of heat capacity on temperature calculated for staphylococcal nuclease at different pH values are presented in Fig. 6.2 by solid lines. The results of experimental measurements form Ref. [12] are presented by symbols. From Fig. 6.2 it is seen that staphylococcal nuclease experience two folding $\leftrightarrow$ transitions in the range of pH between 3.78 and 7.0. At the pH value 3.23 no peaks in the heat capacity is present. It means that the protein exists in the unfolded state over the whole range of experimentally accessible temperatures.

Comparison of the theoretical results with experimental data shows that the theoretical model reproduces experimental behavior better for the solvents with higher pH. The heat capacity peak arising at higher temperatures due to the standard folding $\leftrightarrow$ unfolding transition is reproduced very well for pH values being in the region 4.5–7.0. The deviations at low temperatures can be attributed to the inaccuracy of the statistical mechanics model of water in the vicinity of the freezing point.

The accuracy of the statistical mechanics model for low pH values around 3.88 is also quite reasonable. The deviation of theoretical curves from experimental ones likely arise due to the alteration of the solvent properties at high concentration of protons or due to the change of partial charge of amino acids at pH values being far from the physiological conditions.

Despite some difference between the predictions of the developed model and the experimental results arising at certain temperatures and values of pH the overall performance of the model can be considered as extremely good for such a complex process as structural folding transition of a large biological molecule.

Table 6.3 Values of E_0 for metmyoglobin at different values of solvent pH

pH value	4.10	3.70	3.84	3.5
E_0 (kcal/mol)	1.128	1.150	1.165	1.2

6.3.2 Heat Capacity of Metmyoglobin

Metmyoglobin is an oxidized form of a protein myoglobin. This is a monomeric protein containing a single five-coordinate heme whose function is to reversibly form a dioxygen adduct [25]. Metmyolobin consists of 153 amino acids and it's structure is shown in Fig. 6.1 on the right.

In order to calculate SASA of side chain radicals of metmyoglobin exactly the same procedure as for staphylococcal nuclease was performed (see discussion in the previous subsection). SASA in the folded and unfolded states of the protein has been calculated and is equal 6,847 and 16,926 $\mathring{A}^2$ respectively. Thus, there are 984 H_2O molecules interacting with protein's hydrophobic surface in its unfolded state.

The electrostatic interaction of water molecules with metmyoglobin was accounted for in the same way as for staphylococcal nuclease. The parameter α in (6.15) was chosen to be equal to 2.5. 10,950 H_2O molecules involve in the interaction with the electrostatic field of metmyoglobin in its folded state. The strength of the field was chosen the same as for staphylococcal nuclease.

The parameter κ for metmyoglobin in (6.4), describing the cooperativity of the folding $\leftrightarrow$ unfolding transition, differs significantly from that for staphylococcal nuclease. The transition in metmyoglobin is less cooperative than the transition in staphylococcal nuclease because metmyoglobin has intermediate partially folded states [26]. Thus, while the rigid native-like core of the protein is formed, a significant fraction of amino acids in the flexible regions of the protein can exist in the unfolded state. 1/3 of metmyoglobin's amino acids can considered as being in flexible regions of the protein, i.e. the parameter κ in (6.4) equal to 50.

The values of E_0 in (6.6) differ from that for staphylococcal nuclease and are compiled in Table 6.3. In the calculations for metmyoglobin the values of $A = 1.6\,\mathrm{J\,K^{-1}\,g^{-1}}$, $B = 8.25 \times 10^{-3}\,\mathrm{J\,K^{-2}\,g^{-1}}$ and $T_0 = 323\,\mathrm{K}$ were adjusted in (6.25).

Solid lines in Fig. 6.3 show the dependence of the metmyoglobin's heat capacity on temperature calculated using the developed theoretical model. The experimental data from Ref. [13] are shown by symbols.

Metmyoglobin experiences two folding$\leftrightarrow$ unfolding transitions at the pH values exceeding 3.5 which can be called as cold and heat denaturations of the protein. The dependence of the heat capacity on temperature therefore has two characteristic peaks, as seen in Fig. 6.3. Figure 6.3 shows that at pH lower than 3.84 metmyoglobin exists only in the unfolded state.

The comparison of predictions of the developed theoretical model with the experimental data on heat capacity shows that the theoretical model is well applicable for metmyoglobin case as well. The good agreement of the theoretical and experimental heat capacity profiles over the whole range of temperatures and pH values shows that the model treats correctly the thermodynamics of the protein folding process.

Fig. 6.3 Dependencies of
the heat capacity on
temperature for horse heart
metmyoglobin (see Fig.
6.1b) at different values of
pH. *Solid lines* show the
results of the calculation.
Symbols present the
experimental data from
Ref. [13] The figure is
adopted from [8]

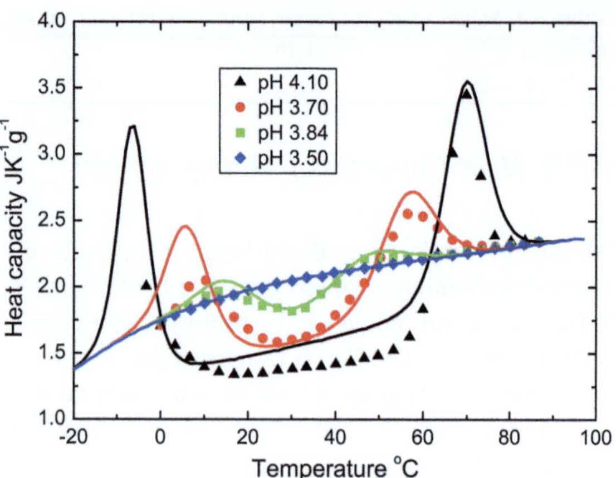

The presented theory includes a number of parameters, namely the energy difference between two phases E_0, strength of the electrostatic field E, number of interacting H_2O molecules α, the parameter describing the cooperativity of the phase transition κ, as well as other parameters introduced in Ref. [10] to treat the partition function of water. Three parameters, E, E_0 and κ, are dependent on the properties of a particular protein and on the pH of the solvent. the values of these parameters were adjusted in order to reproduce the experimental data. All other parameters of the model describing the structure of energy levels of water molecules, their vibrational and librational frequencies, etc. are considered as fixed, being universal for all proteins.

In spite of the model features of the presented approach, let us stress that the complex behavior and the peculiarities in dependencies of the heat capacity on temperature are all very well reproduced by the developed model with only a few parameters. This was demonstrated for two proteins and this result can be considered as a significant achievement. This fact supports the conclusion that the developed model can be used for the prediction of new features of phase transitions in various biomolecular systems. Indeed, from Figs. 6.2 and 6.3 one can extract a lot of useful information on the heat capacity profiles: the concave bending of the heat capacity profile for a completely unfolded protein, the temperature of the cold and heat denaturation, the absolute values of the heat capacity at the phase transition temperature, the broadening of heat capacity peaks. Another peculiarity which is well reproduced by the statistical mechanics model is the decrease of the heat capacity of the folded state of the protein in comparison with that for unfolded state and asymmetry of the heat capacity peaks.

References

1. Muñoz, V. (2007). Conformational dynamics and ensembles in protein folding. *Annual Review of Biophysics and Biomolecular Structure, 36*, 395–412.
2. Shakhnovich, E. (2006). Protein folding thermodynamics and dynamics: Where physics, chemistry, and biology meet. *Chemistry Reviews, 106*, 1559–1588.
3. Dill, K. A., Ozkan, S. B., Shell, M. S, & Weikl, T. R. (2008). The protein folding problem. *Annual Review of Biophysics, 37*, 289–316.
4. Onuchic, J. N., & Wolynes, P. G. (2004). Theory of protein folding. *Current Opinion in Structural Biology, 14*, 70–75.
5. Prabhu, N. V., & Sharp, K. A. (2006). Protein–solvent interactions. *Chemistry Reviews, 106*, 1616–1623.
6. Noetling, B., & Agard, D. A. (2008). How general is the nucleation–condensation mechanism? *Proteins, 73*, 754–764.
7. Yakubovich, A., Solov'yov, A., & Greiner, W. (2010). Conformational changes in polypeptides and proteins. *International Journal of Quantum Chemistry, 110*, 257–269.
8. Yakubovich, A., Solov'yov, A., & Greiner, W. (2009). Statistical mechanics model for protein folding. *AIP Conference Proceedings, 1197*, 186–200.
9. Kumar, S., Tsai, C.-J., & Nussinov, R. (2002). Maximal stabilities of reversible two-state proteins. *Biochemistry, 41*, 5359–5374.
10. Griffith, J. H., & Scheraga, H. (2004). Statistical thermodynamics of aqueous solutions. i. Water structure, solutions with non-polar solutes, and hydrophobic interactions. *Journal of Molecular Structure, 682*, 97–113.
11. Bakk, A., Hoye, J. S., & Hansen, A. (2002). Apolar and polar solvation thermodynamics related to the protein unfolding process. *Biophysical Journal, 82*, 713–719.
12. Griko, Y., Privalov, P., Aturtevant, J., & Venyaminov, S. (1988). Cold denaturation of staphyloccocal nuclease. *Proceedings of the National Academy of Sciences of the United States of America, 85*, 3343–3347.
13. Privalov, P. (1997). Thermodynamics of protein folding. *Journal of Chemical Thermodynamics, 29*, 447–474.
14. Krimm, S., & Bandekar, J. (1980). Vibrational analysis of peptides, polypeptides, and proteins. v. Normal vibrations of β-turns. *Biopolymers, 19*, 1–29.
15. Cubrovic, M., Obolensky, O., & Solov'yov, A. (2009). Semistiff polymer model of unfolded proteins and its application to NMR residual dipolar couplings. *European Physical Journal D, 51*, 41–49.
16. Finkelstein, A., & Ptitsyn, O. (2002). *Protein Physics: A course of Lectures*. Oxford: Elsevier Books.
17. Makhatadze, G., & Privalov, P. (1993). Contribution to hydration to protein folding thermodynamics. i. The enthalpy of hydration. *Journal of Molecular Biology, 232*, 639–659.
18. Cotton, F. A., Edward, J., Hazen, E., & Legg, M. J. (1979). Staphylococcal nuclease: Proposed mechanism of action based on structure of enzyme-thymidine 3',5'-bisphosphate-calcium ion complex at 1.5-a resolution. *Proceedings of the National Academy of Sciences of the United States of America, 76*, 2551–2555.
19. Protein data bank. (2010). http://www.rcsb.org/
20. MacKerell, A., Bashford, D., Bellott, M., Dunbrack, R., Evanseck, J., Field, M., et al. (1998). All-atom empirical potential for molecular modeling and dynamics studies of proteins. *Journal of Physical Chemistry B, 102*, 3586–3616.
21. Phillips, J. C., Braun, R., Wang, W., et al. (2005). Scalable molecular dynamics with NAMD. *Journal of Computational Chemistry, 26*, 1781–1802.
22. Russel, W., Saville, D., & Schowalter, W. (1989). *Colloidal dispersions*. Cambridge: Cambridge University Press.
23. Mallik, B., & Masunov, T. L. A. (2002). Distance and exposure dependent effective dielectric function. *Journal of Computational Chemistry, 23*, 1090–1099.

24. Zhou, H.-X. (2002). Residual charge interactions in unfolded staphylococcal nuclease can be explained by the Gaussian-chain model. *Biophysical Journal, 83*, 2981–2986.
25. Collman, J. P., Boulatov, R., Sunderland, C. J., & Fu, L. (2004). Functional analogues of cytochrome c oxidase, myoglobin, and hemoglobin. *Chemistry Reviews, 104*, 561–588.
26. Shortle, D., & Ackerman, M. S. (2001). Persistence of native-like topology in a denatured protein in 8 m urea. *Science, 293*, 487–489.
27. Chen, J., Lu, Z., Sakon, J., & Stites, W. (2000). Increasing the thermostability of staphylococcal nuclease: Implications for the origin of protein thermostability. *Journal of Molecular Biology, 303*, 125–130.
28. Evans, S., & Brayer, G. (1990). High-resolution study of the three-dimensional structure of horse heart metmyoglobin. *Journal of Molecular Biology, 213*, 885–897.
29. Humphrey, W., Dalke, A., & Schulten, K. (1996). Vmd—visual molecular dynamics. *Journal of Molecular Graphics, 14*, 33–38.

Chapter 7
Summary and Conclusions

This work is devoted to the theoretical study of the conformational transitions in polypeptides and proteins using the methods of statistical physics. The conformational transitions or *folding* of polypeptides and proteins are treated as phase transitions in a finite system. The thesis begins with the analysis of the potential energy surfaces of alanine and glycine tri- and hexapeptides calculated using the ab initio methods of the Density Functional Theory. Analysis of the potential energy surfaces of the polypeptides allowed to distinguish two separate types of degrees of freedom in the system: "hard" and "soft" degrees of freedom, corresponding to the vibrations along covalent bonds in the polypeptide and to twisting along the polypeptide's backbone, respectively. This separation provided the possibility to construct the partition function of a polypeptide experiencing helix $\leftrightarrow$ coil phase transition knowing its potential energy surface as a function of only twisting degrees of freedom.

In Chap. 4 of the thesis a novel ab initio theoretical method for treating the α-helix $\leftrightarrow$ random coil phase transition in polypeptide chains is introduced. The suggested method is based on the construction of a parameter-free partition function for a system undergoing a first order phase transition. All the necessary information for the construction of such a partition function can be calculated on the basis of ab initio DFT, combined with molecular mechanics theories. The suggested method is considered as an efficient alternative to the existing theoretical approaches for the study of helix-coil transition in polypeptides since it does not contain any model parameters. It gives a universal recipe for statistical mechanics description of complex molecular systems. The partition function of polypeptide is written with a minimum number of assumptions about the system which makes the presented theoretical method much more general and universal in comparison with other theoretical approaches.

The detailed analysis of the α-helix $\leftrightarrow$ random coil transition in alanine polypeptides of different lengths is performed in Chap. 5. The potential energy surfaces of polypeptides were calculated with respect to their twisting degrees of freedom and a parameter-free partition function of the polypeptide was constructed. From this partition function, the temperature dependencies of the heat capacity, latent heat and helicity of alanine polypeptides consisting of 21, 30, 40, 50 and 100 amino acids were

A. V. Yakubovich, *Theory of Phase Transitions in Polypeptides and Proteins*,
Springer Theses, DOI: 10.1007/978-3-642-22592-5_7,
© Springer-Verlag Berlin Heidelberg 2011

derived and analyzed. Alternatively, the same thermodynamical characteristics were obtained with the use of molecular dynamics simulations. The results of molecular dynamics simulations were compared with the results of the statistical mechanics approach. The comparison proved the validity of the presented method and established its accuracy. It was demonstrated that the heat capacity of alanine polypeptides has a peak at a certain temperature. The parameters of this peak (i.e. the maximal value of the heat capacity, the temperature of the peak, the width at half maximum, the area of the peak) were analyzed as a function of polypeptide length. Based on the predictions of the two energy-level model, it was demonstrated that the α-helix$\leftrightarrow$ random coil transition in alanine polypeptide is a first order phase transition.

In the work is established the correspondence of the developed method with the results of the semiempirical approach suggested by Zimm and Bragg [1]. For this purpose the key parameters of the Zimm–Bragg semiempirical statistical theory were determined. The calculated parameters of the Zimm–Bragg theory were compared with the results of earlier calculations from reference [2].

The second part of Chap. 5 is devoted to the study of the helix$\leftrightarrow$ random coil transition in alanine, valine, and leucine polypeptides consisting of 30 amino acids in vacuo using the Langevin molecular dynamics approach. It was shown that in the course of the helix$\leftrightarrow$ random coil phase transition the polypeptide chain can experience several sequential structural changes leading to the emergence of additional peaks in the temperature dependence of the heat capacity. This was illustrated by the example of the Val_{30} polypeptide. In this polypeptide, the most prominent heat capacity peak is associated with the breakage of the π-helix secondary structure, while the smaller peak is associated with the structural transition in side chain radicals. The heat capacities of the polypeptides were calculated by two different methods, namely as the derivative of the internal energy with respect to temperature and on the basis of the energy fluctuations in the system. The results obtained by these methods were compared and their accuracy and convergence were established. This analysis revealed that the method based on energy fluctuations can be considered as more general and is, therefore, preferable in most cases. Analysis of the dependence of the accuracy of the heat capacity calculation on duration of MD simulations shows that the accuracy of the simulations is nearly proportional to the simulation time. Although the Chap. 5 is devoted the study of three particular polypeptides, the discussed methods are general and can be applied to the study of more complex systems, such as larger polypeptide chains, proteins, DNA molecules, etc. The suggested formalism can be also applied for the analysis of folding dynamics of polypeptide chains (and proteins) in solution.

In the Chap. 6 of the thesis a novel statistical mechanics model for the description of folding$\leftrightarrow$ unfolding processes in globular proteins is presented. The model is based on the construction of the partition function of the system as a sum over all statistically significant conformational states of a protein. The partition function of each state is a product of partition function of a protein in a given conformational state, partition function of water molecules in pure water and a partition function of H_2O molecules interacting with the protein.

The introduced model includes a number of parameters responsible for certain physical properties of the system. The parameters were obtained from available experimental data and three of them (energy difference between two phases, cooperativity of the transition and the average strength of the protein's electrostatic field) were considered as being variable depending on a particular protein and on pH of the solvent.

The predictions of the developed model were compared with the results of experimental measurements of the dependence of the heat capacity on temperature for staphylococcal nuclease and metmyoglobin. The experimental results were obtained at various pH of the solvent. The suggested model is capable of reproducing well within a single framework a large number of peculiarities of the heat capacity profile, such as the temperatures of cold and heat denaturations, the corresponding maximum values of the heat capacities, the temperature range of the cold and heat denaturation transitions, the difference between heat capacities of the folded and unfolded states of the protein.

The good agreement of the results of calculations obtained using the developed formalism with the results of experimental measurements demonstrates that it can be used for the analysis of thermodynamical properties of many biomolecular systems. Further development of the model can be focused on its advance and application for the description of the influence of mutations on protein stability, analysis of assembly and stability of protein complexes, protein crystallization process, etc.

References

1. Zimm, B., Bragg, J. (1959). Theory of the phase transition between helix and random coil in polypeptide chains. *Journal of Chemical Physics, 31*, 526–535
2. Go, M., Go, N., Scheraga, H. A. (1970). Molecular theory of the helix-coil transition in polyamino acids II. numerical evaluation of s and σ for polyglycine and poly-l-alanine in the absence (for s and σ) and presence (for σ) of solvent. *Journal of Chemical Physics, 52*, 2060–2079

FSC
www.fsc.org
MIX
Papier aus verantwortungsvollen Quellen
Paper from responsible sources
FSC® C105338